Geometry Problem Solving for Middle School

Solutions Manual

Kevin Wang
Kelly Ren
John Lensmire
Wilson Cheung

ISBN: 1-944863-06-0
ISBN-13: 978-1-944863-06-7

First printing, October 2016.

Contents

Introduction

Do you know what geometry is? If you are in middle school, you may have some idea that it has to do with shapes and you are not far off in assuming so. Truth be told, we all possess some kind of "geometric intuition" even before we take this class subject in school. "Geometric intuition" refers to the fact that when we look at the world around us, we instantly recognize patterns found in nature. Everything, including the sun, plants and the rest of the world around us, can all be seen in terms of mathematics and the shapes found within these can be understood in the context of geometry. If it were not for mathematics, for example, we may not have known that the intricately curved design of the Nautilus shell actually applies the concept of the Fibonacci sequence to perfection.

Geometry is also an essential part of many people's lives especially when it comes to their work in a specific career. Architectual design might be one of the first application fields of geometry that comes to mind. Architects employ geometric concepts for the practical design of a structure (such as determining the angle from the structure to the ground that will allow for a certain height and weight distribution) to the aesthetic look of a structure (i.e. why no two buildings are alike). Modern development in STEAM (Science, Technology, Engineering, Arts, and Mathematics) fields further stresses the importance of mathematics. Geometry is one of the essential subject areas to lay the foundations for students to eventually excel in trending fields such as robotics design,

3D animation, VR (the fascinating world of virtual reality), GPS (Global Positioning System), aeronautical and aerospace engineering (for flight and space vehicles such as those who have designed and currently run the International Space Station) and for innovative automotive engineering (such as autonamous cars to run the way they are designed). Math therefore provides a way for people to make sense of the world!

Although careers may not be your immediate concern, geometry does provide a foundation in problem solving that will develop over time. Geometry provides a natural way to transition from "geometric intuition" to abstract thinking. In our efforts to assist students, this curriculum book provides an invaluable introduction to the mathematical concepts found in geometry. The content of this book presents geometric concepts in an easily accessible way for students in middle school to understand while providing the kind of academic challenge they need to help them improve their critical thinking skills. In this way, the intrinsic "geometric intuition" that students already possess is molded to allow for more abstract thinking, necessary for understanding geometry.

To accomplish this feat, students will find that the problems contained within these pages are related to situations found in their daily lives. By using real-life scenarios to demonstrate mathematics, students will improve their spatial visualization, helping them easily transition to the kind of problem solving that is needed when completing, for instance, a more formal geometric proof. Geometry is a standard high school mathematics course serving as a gateway to more advanced mathematics later in high school and at the collegiate level. Earlier exposure to the basic ideas and vocabulary used in geometry, will allow students to be confident in approaching geometry again when taken as a subject in school.

In this book, we present more in-depth problem solving, covering the application of fundamental concepts in areas, angles, surface areas and volumes and how students can readily apply these concepts in their own lives, highlighted with pictures and 3D shapes to illustrate the problems. The book covers in-depth implementation of Common Core Math Standards for geometry that all middle school students are required to understand before entering high school.

Finally, this book serves as a training tool for students interested in participating in and expanding upon their knowledge of middle school math competitions. Among others, the materials covered in the book have a proven record of effectiveness for those students who want to excel in the American Math Competition (AMC) 8, MathCounts, and the ZIML Math Competitions for Division M hosted monthly. Major competition problems are reviewed and explained, allowing for students to realize the multiple ways in which they can maneuver through math problems with ease. Math equations are shown to be multi-layered and as such a variety of methods must be employed to solve every problem and come to a complete solution. As a result, students improve their problem solving techniques and benefit from their newfound skills in thinking more critically when approaching competition mathematics.

It is our hope at Areteem Institute that you will find this curriculum book helpful in providing a solid foundation at the middle school level for the success of studying high school geometry and beyond. Students will be able to approach geometry with a better understanding of its concepts and build the confidence to solve problems, whether in a school class setting or in the heat of the moment of a math competition. This book serves as a Solutions Manual for the Student Workbook, titled "*Geometry Problem Solving for Middle School*", which contains key concepts, examples and their solutions, and problem solving strategies. We hope that you enjoy this book!

Common Core and This Book

Teachers and students that are in 6th, 7th, and 8th grade math can use this book to teach and learn mathematical reasoning and problem solving, focusing on concepts in the Common Core Geometry domain.

For reference, a summary of the middle school geometry domain is provided below.

Standard(s)	Cluster
6.G.1-4	Solve real-world and mathematical problems involving area, surface area, and volume.
7.G.1-3	Draw, construct, and describe geometrical figures and describe the relationships between them.
7.G.4-6	Solve real-life and mathematical problems involving angle measure, area, surface area, and volume.
8.G.1-5	Understand congruence and similarity using physical models, transparencies, or geometry software.
8.G.6-8	Understand and apply the Pythagorean Theorem.
8.G.9	Solve real-world and mathematical problems involving volume of cylinders, cones, and spheres.

The start of each chapter summarizes the specific Common Core standards emphasized in the chapter. While the focus of the book is geometry, the problem solving stressed in the exercises allows students to practice in other domains as well. Exercises with calculations allow students to put the standards taught in the "Number System" (NS) and "Expressions and Equations" (EE) domains to good use. Geometric concepts such as similarity also provide excellent examples for use in the "Ratios and Proportional Relationships" (RP) domain.

For more details about the specific standards, clusters, and domains quoted above, see www.corestandards.org/Math where the full Mathematical Standards are available for download.

About Areteem Institute

Areteem Institute is an educational institution that develops and provides in-depth and advanced math and science programs for K-12 (Elementary School, Middle School, and High School) students and teachers. Areteem programs are accredited supplementary programs by the Western Association of Schools and Colleges (WASC). Students may attend the Areteem Institute through these options:

- Live and real-time face-to-face online classes with audio, video, interactive online whiteboard, and text chatting capabilities;
- Self-paced classes by watching the recordings of the live classes;
- Topical short video courses for trending math, science, technology, engineering, English, and social studies topics;
- Summer Intensive Camps on prestigious university campuses and Winter Boot Camps;
- Practice with selected daily problems for free, monthly ZIML competition at `ziml.areteem.org`.

The Areteem courses are designed and developed by educational experts and industry professionals to bring real world applications into STEM education. The programs are ideal for students who wish to build their mathematical strength in order to excel academically and eventually win in Math Competitions (AMC, AIME, USAMO, IMO, ARML, MathCounts, Math Olympiad, ZIML, and other math leagues and tournaments, etc.), Science Fairs (County Science Fairs, State Science Fairs, national programs like Intel Science and Engineering Fair, etc.) and Science Olympiad, or purely want to enrich their academic lives by taking more challenges and developing outstanding analytical, logical thinking and creative problem solving skills.

Since 2004 Areteem Institute has been teaching with methodology that is highly promoted by the new Common Core State Standards: stressing the conceptual level understanding of the math concepts, problem solving techniques, and solving problems with real world applications. With the guidance from experienced and passionate professors, students are motivated to explore concepts deeper by identifying an interesting problem, researching it, analyzing it, and using a critical thinking approach to come up with multiple solutions.

Thousands of math students who have been trained at Areteem achieved top honors and earned top awards in major national and international math competitions, including Gold Medalists in the International Math Olympiad (IMO), top winners and qualifiers at the USA Math Olympiad (USAMO/JMO), and AIME, top winners at the Zoom

International Math League (ZIML), and top winners at the MathCounts National. Many Areteem Alumni have graduated from high school and gone on to enter their dream colleges such as MIT, Cal Tech, Harvard, Stanford, Yale, Princeton, U Penn, Harvey Mudd College, UC Berkeley, UCLA, etc. Those who have graduated from colleges are now playing important roles in their fields of endeavor.

Further information about Areteem Institute, as well as updates and errata of this book, can be found online at http://www.areteem.org.

Acknowledgments

This book contains many years of collaborative work by the staff of Areteem Institute. This book could not have existed without their efforts. Huge thanks go to the Areteem staff for their contributions, in particular, Cameron Yanoscik for his help with drafting the overview of the introduction and providing editorial update of the book.

The examples and problems in this book were either created by the Areteem staff or adapted from various sources, including other books and online resources. Especially, some good problems from American Mathematics Competitions (AMC) 8 and MATH-COUNTS are chosen as examples to illustrate concepts or problem-solving techniques. The original resources are credited whenever possible. However, it is not practical to list all such resources. We extend our gratitude to the original authors of all these resources.

1. Counting Through Patterns

Problem 1.1 The diagram below is used for the following four questions.

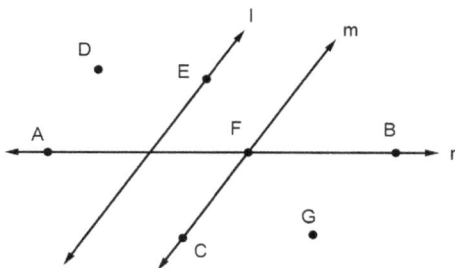

How many points are there in the picture? How many lines?

Answer

7 points; 3 lines

Problem 1.2 What points are on line n? Are there any points that lie on two lines? Name a point not on any line.

Answer

A, B, F; F; D or G

Problem 1.3 Which line is determined by the points C and F? That is, what is another name for the line $\overleftrightarrow{CF}$?

Answer

m

Problem 1.4 What is another way to describe or determine line n?

Answer

$\overleftrightarrow{AB}, \overleftrightarrow{AF}, \overleftrightarrow{BF}$

Problem 1.5 How many triangles are in each of the diagrams below?

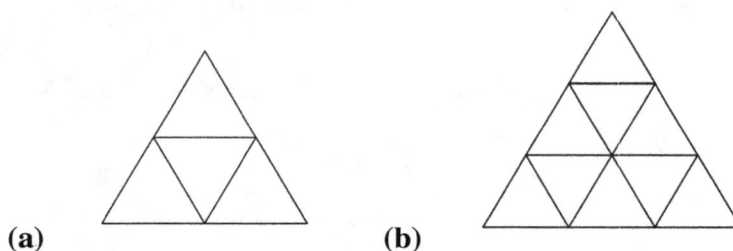

(a)   (b)

Answer

a) 5; b) 13

Solution

Count by sizes:

For (a), note that there are 4 smaller triangles and 1 larger triangle. This yields a total of

$$4 + 1 = 5$$

---

triangles.

For (b), note that there are 9 smaller triangles, 3 medium triangles, and 1 big triangle. This yields a total of

$$9 + 3 + 1 = 13$$

triangles.

Problem 1.6 **In the following diagram, $\angle 1 = \angle 2 = \angle 3$. The sum of the measures of all possible angles in $\angle AOB$ is $180°$. What is the measure of $\angle AOB$?**

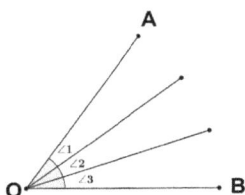

Answer

$54°$

Solution

There are 6 angles in total. When adding up all the angles, $\angle 1$ and $\angle 3$ are counted 3 times each, and $\angle 2$ is counted 4 times. Therefore

$$3\angle 1 + 4\angle 2 + 3\angle 3 = 180°.$$

Since

$$\angle 1 = \angle 2 = \angle 3,$$

we get

$$10\angle 1 = 180°,$$

thus

$$\angle 1 = 180 \div 10 = 18°.$$

So

$$\angle AOB = 3\angle 1 = 3(18) = 54°.$$

Problem 1.7 In the following diagram, each small equilateral triangle has area 1. Find the total area of all the triangles in the figure below.

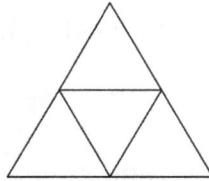

Answer

  8

Solution

  In the diagram, there are 4 small triangles and 1 large triangle. Each small equilateral triangle has area of 1. Since the large triangle is made up of 4 small triangles, the area of the large triangle is 4. Therefore, the total area of all triangles is

$$4 \times 1 + 1 \times 4 = 8.$$

Problem 1.8 How many squares are there in the following diagram?

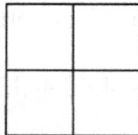

Answer

  5

Solution

There are 4 small squares and 1 big square. The total number of squares is

$$4 + 1 = 5.$$

Problem 1.9 **How many angles less than $180°$ are there in the following diagram?**

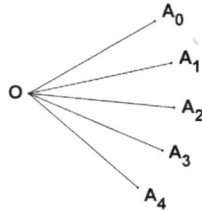

10

There are 4 small angles

$$\angle A_0OA_1, \angle A_1OA_2, \angle A_2OA_3, \angle A_3OA_4,$$

3 medium angles

$$\angle A_0OA_2, \angle A_1OA_3, \angle A_2OA_4,$$

and 2 large angles

$$\angle A_0OA_3, \angle A_1OA_4,$$

and 1 huge angle

$$\angle A_0OA_4.$$

Therefore, the number of angles in the diagram is

$$4 + 3 + 2 + 1 = 10.$$

Problem 1.10 **In the following diagram, $\angle 1 = 3\angle 3, \angle 2 = 2\angle 3$. The sum of the measures of all the angles is $180°$. What is the measure of $\angle AOB$?**

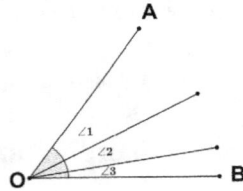

Answer

54°

Solution

There are 6 angles in total. When adding up all the angles, $\angle 1$ and $\angle 3$ are counted 3 times each, and $\angle 2$ is counted 4 times. Therefore

$$3\angle 1 + 4\angle 2 + 3\angle 3 = 180°.$$

Since $\angle 1 = 3\angle 3$ and $\angle 2 = 2\angle 3$, we get

$$(3 \times 3 + 2 \times 4 + 1 \times 3)\angle 3 = 20\angle 3 = 180°,$$

thus $\angle 3 = 180 \div 20 = 9°$. So

$$\angle AOB = \angle 1 + \angle 2 + \angle 3 = 6\angle 3 = 54°.$$

Problem 1.11 **Count the triangles in each of the following diagrams:**

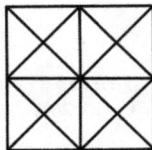   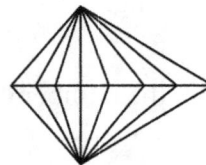

a)   **b)**

Answer

a) 44, b) 63

Solution

To count accurately, categorize the triangles.

In a), there are 16 one-piece triangles , 16 two-piece triangles, 8 four-piece triangles, and 4 eight-piece triangles for a total of

$$16 + 16 + 8 + 4 = 44$$

triangles.

In b), the top half of the diagram has

$$7 + 6 + 5 + 4 + 3 + 2 + 1 = 28$$

triangles. Similarly the bottom has 28. Lastly there are 7 triangles which overlap the top and bottom for a total of 63 triangles.

Problem 1.12 Each small equilateral triangle has area 1 in the diagram below. Find the total area of all the triangles.

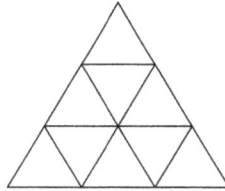

Answer

30

Solution

In the diagram, there are 9 small triangles, 3 medium triangles and 1 large triangle. Each small equilateral triangle has area of 1. Since the medium triangle is made up of 4 small triangles, the area of the medium triangle is 4. Similarly, since the large triangle is made up of 9 small triangles, the area of the large triangle is 9. Therefore, the total area of all triangles is

$$9 \times 1 + 3 \times 4 + 1 \times 9 = 30.$$

Problem 1.13 **How many squares are there in the following diagram?**

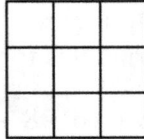

Answer

14

Solution

There are 9 small squares, 4 medium squares, and 1 big square. Adding there are

$$9 + 4 + 1 = 14$$

squares in total.

Problem 1.14 (2008 AMC 8) In the figure, what is the ratio of the area of the gray squares to the area of the white squares?

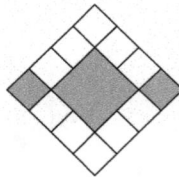

Answer

$3 : 5$

Solution

First, note that the gray square can be divided into four smaller squares. Therefore, there are 6 gray tiles and 10 white tiles, giving a ratio of $3 : 5$.

Problem 1.15 **In the following diagram, $ABCD$ is a parallelogram, and each of the segments in the diagram is parallel to one of $\overline{AB}$, $\overline{AD}$, or $\overline{BE}$. Count the number of parallelograms in the diagram that contain the shaded triangle.**

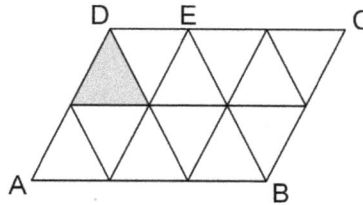

7

If we ignore the bottom half of the diagram, we can create 3 parallelograms (made up of 2, 4, and 6 triangles). Including the bottom half creates an additional 3 parallelograms (made up of 4, 8, and 12 triangles). Lastly, if we take the shaded region and join it with the triangle below, we create 1 parallelogram. There are

$$3 + 3 + 1 = 7$$

parallelograms containing the shaded region in the diagram.

Problem 1.16 **Arrange several rhombi, all of whose side lengths are 2cm, to form a long parallelogram, as shown in the diagram. Assume the perimeter of the long parallelogram is 144cm, how many rhombi are there?**

35

Solution

The left and right sides each has length 2 cm, so the remaining sides of the parallelogram have length 140 cm. Therefore, the length of the top of the long parallelogram is

$$140 \div 2 = 70 \text{cm}.$$

As the top of each rhombi is 2 cm, there are

$$70 \div 2 = 35$$

rhombi.

Problem 1.17 **Given a circular disk, use 6 lines to divide the disk into small regions. At most how many regions can there be?**

Answer

22

Solution

One line makes 2 regions. Two lines make 4 regions.

Suppose n lines are already present, and the $(n+1)$-th line should intersect with all the existing lines, and the $(n+1)$-th line itself is divided into n pieces, and each piece create a new region. Therefore the answer is

$$2 + 2 + 3 + 4 + 5 + 6 = 22.$$

Problem 1.18 **(2002 AMC 8) A circle and two distinct lines are drawn on a sheet of paper. What is the largest possible number of points of intersection of these figures?**

Answer

5

Solution

Note that the maximum number of points a line can intersect a circle is 2.

Additionally, the lines can intersect each other. This yields a maximum total of

$$2 + 2 + 1 = 5$$

intersections.

Problem 1.19 **In the diagram, each side is perpendicular to its adjacent sides, and all small sides have equal length. Given that the perimeter of this diagram is** 100, **find the area of the shape.**

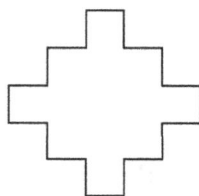

Answer

325 cm^2

Solution

Note that the diagram has the same perimeter as a square with the same dimensions. The square thus has a side length of

$$100 \div 4 = 25$$

cm. As a side of the square is made up of 5 small sides from the diagram, each small side has length

$$25 \div 5 = 5$$

cm. As the diagram is made up of

$$1 + 3 + 5 + 3 + 1 = 13$$

squares with this side length, the total area of the shape is

$$13 \times 5^2 = 325 \text{ cm}^2.$$

Problem 1.20 Use 4 congruent rectangles to form one big square, as shown. The big square has area 100 cm^2. Suppose the side length of the small square in the center is 2cm. What is the perimeter of each rectangle?

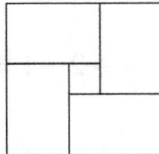

Answer

20 cm

Solution

Note that the side length of the big square is equal to the sum of the width and length of the rectangle. Since the length and width differ by 2 and must add up to 10, the length of the rectangle is 6 and the width of the rectangle is 4. Therefore, the perimeter of one rectangle is

$$2 \times 6 + 2 \times 4 = 20\text{cm}.$$

Problem 1.21 (2002 AMC 8) A corner of a tiled floor is shown. If the entire floor is tiled in this way and each of the four corners looks like this one, then what fraction of the tiled floor is made of darker tiles?

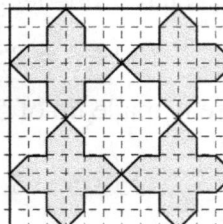

Answer

$\dfrac{4}{9}$

Solution

Note that the pattern repeats for every 6×6 tile.

Looking closer at the 6×6 tile, there is symmetry of the top 3×3 square, so the fraction of the entire floor in dark tiles is the same as the fraction in the square.

In the 3×3 grid, there are 4 dark tiles, and 9 total tiles and therefore, the answer is $\frac{4}{9}$.

Problem 1.22 **In the diagram, each side is perpendicular to its adjacent sides, and all small sides have equal length. Given that the area of this shape is 117, find the perimeter.**

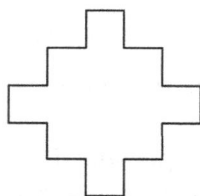

Answer

60 cm

Solution

Note the diagram is made up of

$$1 + 3 + 5 + 3 + 1 = 13$$

identical squares. Thus each square must have area

$$117 \div 13 = 9\text{cm}^2,$$

so each square has side length 3 cm. Now note that all the sides can be rearranged to form a square with side length

$$5 \times 3 = 15$$

cm. Hence the perimeter is

$$4 \times 15 = 60\text{cm}.$$

Problem 1.23 (ZIML 2016) If we cut a circular ring vertically it divides the ring into 2 pieces. A second vertical cut divides the ring into 4 pieces. If we cut the circular ring 10 times vertically, how many pieces do we divide the ring into?

Answer

20

Solution

Observe a pattern. When one cut is made onto the circular ring, there are 2 pieces.

When two cuts are made, there are 4 pieces.

Any additional cut that is made will increase the number of pieces by 2.

Therefore, there are 20 total pieces after 10 cuts are made.

Problem 1.24 Arrange several equilateral triangles and rhombi, all of whose side lengths are 2cm, to form a long parallelogram, as shown in the diagram. Assume the perimeter of the long parallelogram is 244 cm, how many equilateral triangle and rhombi are there?

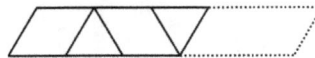

Answer

40 of each

Solution

The left and right sides each has length 2cm, so the top and bottom sides each has length

$$(244 - 2 \times 2) \div 2 = 120$$

---

cm. Each group of 2 rhombi and 2 triangles has top side length 6cm, so there are 20 such groups, and thus there are 40 rhombi and 40 triangles.

Problem 1.25 (ZIML 2016) How many isosceles triangles can be formed using the dots in the following array as vertices? (The dots are evenly spaced.)

Answer

32

Solution

Since the dots are equally spaced, we can create isosceles triangles given as follows:

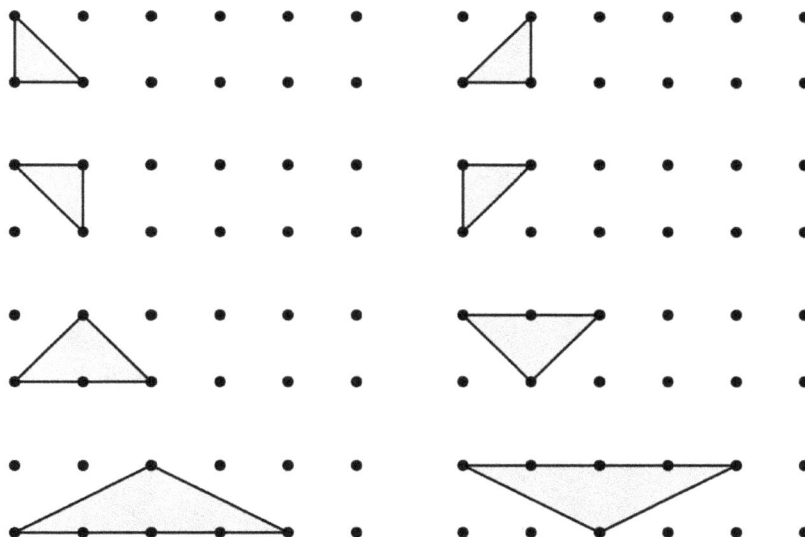

Respectively, there are $5, 5, 5, 5, 4, 4, 2$, and 2 possibilities, yielding a total of 32 isosceles triangles.

2. Practice with Measurements

Problem 2.1 Suppose that the diagram below is not to scale and that the length of AB is 12 and the length of AD is 10. Furthermore, assume that $AD = BE$ and all triangles in the figure below are congruent.

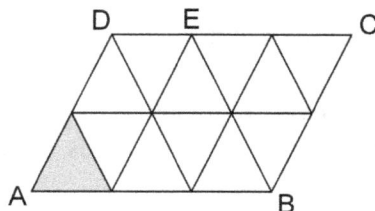

Find the perimeter of parallelogram $ABCD$.

Solution

The perimeter is $AB + BC + CD + AD$. We know $AD = BC$ and $AB = CD$ because $ABCD$ is a parallelogram. Therefore

$$2 \cdot AB + 2 \cdot AD = 2 \cdot 12 + 2 \cdot 10 = 44$$

is the perimeter.

Problem 2.2 **Find the perimeter of the one of the small triangles in the above diagram.**

Answer

    14

Solution

Note that 3 bases of the triangles make up AB so the base of each triangle is

$$12 \div 3 = 4.$$

Similarly, two sides make up AD, so the side of the triangle is

$$10 \div 2 = 5.$$

Hence the perimeter is the base plus the two sides so is $4 + 5 + 5 = 14$.

Problem 2.3 **Suppose I remove the bases of all the small triangles and I want to travel on a zig-zag path from A to B. How long is the path I take?**

Answer

    30

Solution

Note that there are 4 different zig-zag paths, but all of the travel on a total of 6 sides. Therefore the path is of length $6 \cdot 5 = 30$.

Problem 2.4 Suppose that the diagram is not to scale and this time you know that the largest triangle in the diagram (made up of the small triangles) has area 8. (You do not know the lengths AB or CD.)

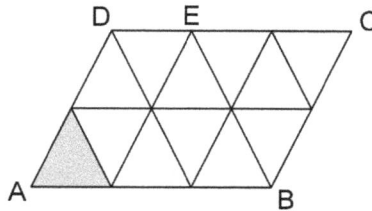

Find the area of a single small triangle.

Answer

> 2

Solution

> The large triangle we can find in the diagram is made up of 4 small triangles, so a small triangle has area $8 \div 4 = 2$.

Problem 2.5 Find the area of the entire parallelogram.

Answer

> 24

Solution

> As calculated above, each small triangle has area 2, so the entire parallelogram, which is made up of 12 triangles, has area $12 \times 2 = 24$.

Problem 2.6 A rectangle is divided into 3 squares, as shown in the diagram. Given that the area of the rectangle is 150 in^2. Find the length and width of the rectangle.

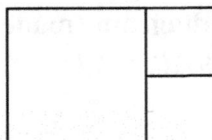

Answer

10 in and 15 in

Solution 1

Let a be the side length of the bigger square, b be the side length of the smaller square, then $a = 2b$, and $a^2 + 2b^2 = 150$. We have $a = 2b$, so $a^2 = 4b^2$, so

$$a^2 + 2b^2 = 4b^2 + 2b^2 = 6b^2 = 150.$$

Hence, $b^2 = 25$ and $b = 5$, so $a = 10$. The width of the rectangle is 10 in, and the length is $10 + 5 = 15$ in.

Solution 2

Note the larger square can be divided into four copies of the smaller square, and thus the rectangle is made up of 6 small squares. Therefore, the small square has area

$$150 \div 6 = 25 \text{ in}^2$$

and has side length 5 in. Therefore the entire rectangle is $5 \times 3 = 15$ by $5 \times 2 = 10$.

Problem 2.7 **A big rectangle is divided into 6 squares of different sizes, as shown. Given that the smallest square in the middle has area 4 cm^2 and the length of the big rectangle is 26, find the area of the big rectangle.**

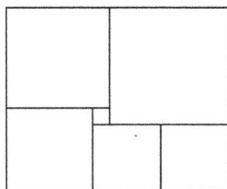

Answer

143 cm^2

Solution

The smallest square has a side length of 2 cm. Let x be the side length of the square at the lower right corner. Then the side lengths of the squares (except for the smallest one), in clockwise order, are $x, x, x+2, x+4, x+6$. If the length of the rectangle is 26, we have

$$x+x+x+2 = 26$$

so $3x = 24$ and $x = 8$. We then see that the rectangle has width

$$x+(x+6) = 8+14 = 22$$

cm. Hence the area is

$$22 \times 26 = 572$$

cm^2.

Problem 2.8 A big rectangle is divided into 7 smaller congruent rectangles, as shown. Given that the area of the big rectangle is 42 cm^2, find the perimeter of the big rectangle.

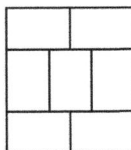

Answer

26 cm

Solution

Each smaller rectangle has area

$$42 \div 7 = 6$$

cm^2. By comparing the top row of 2 horizontal rectangles and the middle row of 3 vertical ones we see that the ratio between the length and width of the smaller rectangle is $3 : 2$. Since $3 \times 2 = 6$, we see that each small rectangle is a 3 cm by 2 cm rectangle. Hence, the entire perimeter is thus

$$6 \times 3 + 4 \times 2 = 26$$

centimeters.

Problem 2.9 **A rectangle is divided into 4 smaller rectangles by two lines, as shown. The perimeters of three of these rectangles are 12, 14, and 14. Find the perimeter of the remaining (shaded) rectangle.**

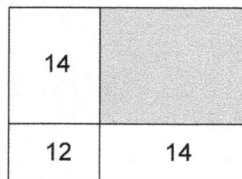

Answer

16

Solution 1

Start with the lower left rectangle. Half the perimeter is 6, so we see that a 4×2 rectangle has perimeter 12. Let's assume the lower left rectangle has those dimensions. Then the length of the upper left rectangle also has length 4. Hence it must have height

$$14 \div 2 - 4 = 7 - 4 = 3.$$

Similarly we see the lower right rectangle then has height 2 and length

$$14 \div 2 - 2 = 7 - 2 = 5.$$

Hence the upper right triangle has dimensions 3 and 5, so

$$2(3 + 5) = 2 \times 8 = 16$$

is its perimeter.

<h2>Solution 2</h2>

Let x denote the length of the lower left rectangle. If its perimeter is 12, the sum of its length and height must be $12 \div 2 = 6$ and thus must have a height of $6 - x$. Similarly, the upper left rectangle has dimensions x by $7 - x$. As the lower right rectangle has the same height as the lower left rectangle, its dimensions must be $x + 1$ by $6 - x$. Therefore, the dimensions of the shaded rectangle are $x + 1$ by $7 - x$ hence its perimeter is

$$2(x + 1 + 7 - x) = 2 \times 8 = 16.$$

Problem 2.10 (2008 AMC 8) In square $ABCE$, $AF = 2FE$ and $CD = 2DE$. What is the ratio of the area of $\triangle BFD$ to the area of square $ABCE$?

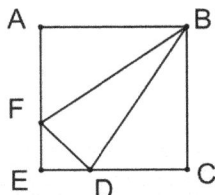

<h2>Answer</h2>

$$\frac{5}{18}$$

Solution

Note that the area of $\triangle BFD$ is the area of square $ABCE$ minus the the area of the three triangles around it. Arbitrarily, let us assign the side length of the square to be 6, so the full square has area $6^2 = 36$. Since $AF = 2FE$ and $CD = 2DE$ we have

$$AF = CD = 4, EF = DE = 2.$$

Hence the area of triangle $\triangle BFD$ is

$$6^2 - \frac{1}{2} \times 6 \times 4 - \frac{1}{2} \times 2 \times 2 - \frac{1}{2} \times 4 \times 6 = 36 - 12 - 2 - 12 = 10.$$

The ratio of areas is then

$$\frac{10}{36} = \frac{5}{18}.$$

Problem 2.11 **The perimeter of rectangle $ABCD$ is 20 cm. Construct a square on the top and right sides of $ABCD$ as shown below. Given that the sum of the areas of these squares is 52 cm^2, find the area of rectangle $ABCD$.**

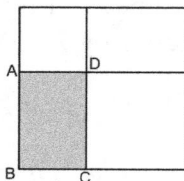

Answer

24 cm^2

Solution

First note that the entire diagram is a square with dimensions 10×10, made up of the two squares and two congruent rectangles (one of which is $ABCD$). Hence the area of the square is

$$100 = 52 + 2[ABCD].$$

Therefore,

$$[ABCD] = (100 - 52) \div 2 = 24$$

cm.

Problem 2.12 Two rectangles and one square are assembled to form a big square as shown. The areas of the rectangles are **44** and **28**. What is the area of the smaller (lower-right) square?

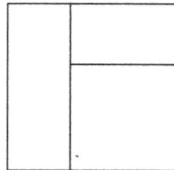

Answer

49

Solution

Extend the side of the smaller rectangle to cut the bigger rectangle and form a tiny square as shown.

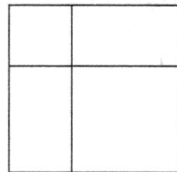

The tiny square has area

$$44 - 28 = 16$$

and hence a side length of 4. The small rectangle thus has dimensions 4×7 and the smaller square is 7 by 7 with area $7^2 = 49$.

Problem 2.13 **A rectangle is divided into 4 squares, as shown in the diagram. Given that the area of the bigger square is 16 in^2 more than one of the smaller squares, find the area of the whole rectangle.**

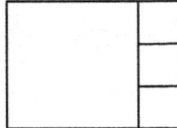

Answer

24 in^2

Solution

Let a be the side length of the bigger square, b be the side length of the smaller square, then $a = 3b$, and since the bigger square is 16 in^2 larger we have

$$a^2 = 9b^2 = b^2 + 16,$$

Thus, $8b^2 = 16$ and $b^2 = 2$, so $a^2 = 18$. The area of the rectangle is $a^2 + 3b^2 = 24$ in^2.

Problem 2.14 **A rectangle is divided into 4 squares, as shown in the diagram. Given that the area of the rectangle is 300 in^2. Find the length and width of the rectangle.**

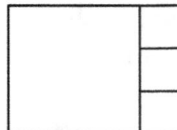

Answer

20 in and 15 in

Solution

Let a be the side length of the bigger square, b be the side length of the smaller square, then $a = 3b$, and since the whole rectangle has area 300 in^2

$$a^2 + 3b^2 = 300.$$

Thus $a = 3b$, and $a^2 = 9b^2$, so $a^2 + 3b^2 = 9b^2 + 3b^2 = 12^2 = 300$. Hence, $b^2 = 25$ and $b = 5$, so $a = 15$. The length of the rectangle is $15 + 5 = 20$ in, and the width is 15 in.

Problem 2.15 (2009 AMC 8) The length of a rectangle is increased by 10% percent and the width is decreased by 10% percent. What percent of the old area is the new area?

Answer

99%

Solution

Suppose the original dimensions of the square are 10×10. (Since we are dealing with a percentage the actual dimensions do not matter.) Note that

$$10\% \times 10 = 1,$$

so the new rectangle has dimensions $10 + 1 = 11$ by $10 - 1 = 9$. The ratio of the old area to the new area is

$$\frac{99}{100} = 0.99 = 99\%.$$

Problem 2.16 A big rectangle is divided into 10 smaller congruent rectangles, as shown. Given that the area of the big rectangle is 120 cm^2, find the perimeter of the big rectangle.

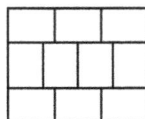

Answer

44 cm

Solution

Each smaller rectangle has area 12 cm^2. If we compare the top row of 3 horizontal rectangles and the middle row of 4 vertical ones we see the ratio between the length and width of the smaller rectangle is 4 : 3.

Notice that $4 \times 3 = 12$ so in fact each small rectangle has dimensions 4×3. Thus the base of the big rectangle is

$$4 + 4 + 4 = 12$$

and the height is

$$3 + 4 + 3 = 10$$

so the total area is $12 \times 10 = 120$ cm.

Problem 2.17 **A rectangle is divided into 4 smaller rectangles by two lines, as shown. The areas of three of these rectangles are 6, 12, and 10. Find the area of the remaining (shaded) rectangle.**

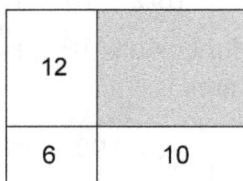

Answer

20

Solution 1

We know a 3×2 rectangle has area 6, so since we are not given dimensions, let's assume the lower left rectangle has base 3 and height 2. From this we see the upper left also has base 3 and height

$$12 \div 3 = 4.$$

Similarly the lower right triangle has dimensions

$$2 \times 5.$$

Hence the upper right triangle has base 5 and height 4, so area $5 \times 4 = 20$.

Solution 2

Let the remaining rectangle have area x. Because the rectangles share either their bases or their heights, areas are proportional:

$$\frac{6}{12} = \frac{10}{x},$$

so cross-multiplying we have $6x = 20$ and $x = 20$, the area of the shaded rectangle.

Problem 2.18 (2011 AMC 8) Two congruent squares, *ABCD* and *PQRS*, have side length 15. They overlap to form the 25 by 15 rectangle *AQRD* shown. What percent of the area of rectangle *AQRD* is shaded?

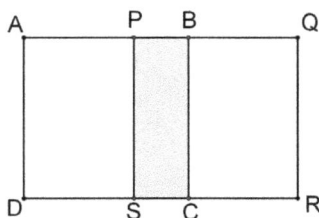

Answer

20%

Solution

If there was no overlap, the squares would form a rectangle with width 15 and length

$$15 + 15 = 30.$$

Therefore the overlap is $30 - 25 = 5$. The width is still 15, so the area is

$$5 \times 15 = 75.$$

The area of the whole shape is $25 \times 15 = 375$, so the shaded rectangle has area

$$\frac{75}{375} = \frac{1}{5} = 0.20 = 20\%$$

of the full rectangle $AQRD$.

Problem 2.19 **The perimeter of rectangle $ABCD$ is 10 cm. Construct a square on the top and right sides of $ABCD$ as shown below. Given that the sum of the areas of these squares is 13 cm^2, find the area of rectangle $ABCD$.**

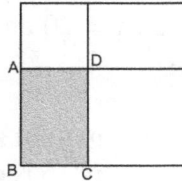

Answer

6 cm^2

Solution

The full diagram is a square, and unfolding BA and AD we see that half the perimeter of $ABCD$ is the side length of this square. Hence this square has side length $10 \div 2 = 5$ and area $5^2 = 25$. This area is made up of the two squares and two congruent rectangles (one of which is $ABCD$). The squares have area 13, so two copies of $ABCD$ has area

$$25 - 13 = 12$$

so $ABCD$ has area $12 \div 2 = 6$.

Problem 2.20 The shape in the diagram consists of 2 congruent squares and 3 congruent rectangles, and its perimeter is 28. If $AB = 2 \times BC$, what is the total area of the diagram?

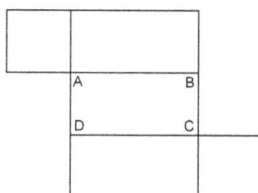

Answer

32

Solution

Since $AB = 2 \times BC$, the rectangles all have a base that is twice the side length of the squares. Since the rectangles' heights are the same as the side length of the square, the perimeter of the shape is 14 times the side length of the square. Hence the side length of the square is 2. Since the full diagram can be exactly covered by

$$1 + 2 + 2 + 2 + 1 = 8$$

of the squares, the total area of the diagram is

$$8 \times 2^2 = 32.$$

Problem 2.21 (2012 AMC 8) A rectangular photograph is placed in a frame that forms a border two inches wide on all sides of the photograph. The photograph measures 8 inches high and 10 inches wide. What is the area of the border, in square inches?

Answer

88

Solution

The area of the frame is equal to the area of the photograph and frame minus the area of the photograph. The height of the whole frame (including the photograph) is

$$8 + 2 + 2 = 12,$$

and the width of the whole frame,

$$10 + 2 + 2 = 14.$$

Therefore, the area of the whole figure would be

$$12 \times 14 = 168$$

square inches. Since the area of the photograph is

$$8 \times 10 = 80$$

square inches, the area of the frame is $168 - 80 = 88$ square inches.

Problem 2.22 **Four congruent rectangles and one square are assembled into one big square. The areas of the two squares are 64 and 16 respectively. What are the length and width of the rectangles?**

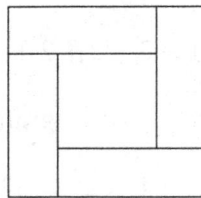

Answer

6 and 2

Solution

The side lengths of the squares are 8 and 4 respectively. Note the big square's side length is equal to the small square's side length plus two widths of the rectangles. Hence the rectangles have width

$$(8-4) \div 2 = 2.$$

Similarly, the side length is also the width plus the length of the rectangles, so

$$8 - 2 = 6$$

is the length of the rectangles.

Problem 2.23 **Divide a big square into 6 congruent rectangles, as shown. Given that each of the rectangles has perimeter 100, find the area of the big square.**

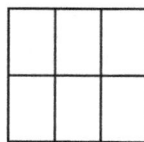

Answer

3600

Solution

Let a be the length of each of the rectangles, and b be the width. The perimeter is 100, so

$$2a + 2b = 100,$$

and since the full shape is a square

$$2a = 3b.$$

Therefore we have

$$3b + 2b = 5b = 100$$

so $b = 100 \div 5 = 20$ and $2a = 100 - 2 \times 20$ so $a = 60 \div 2 = 30$. Therefore the side length of the square is $2a = 60$, and finally the area of the big square is $60^2 = 3600$.

Problem 2.24 (2012 AMC 8) A square with integer side length is cut into 10 squares, all of which have integer side length and at least 8 of which have area 1. What is the smallest possible value of the length of the side of the original square?

Answer

4

Solution

There must be 10 squares with integer side lengths, so the full square must have side length greater than 3. If the side length of the square is 4, the area of the square is 16 and 16 can be partitioned into 8 squares with area 1 and two squares with area 4. Since the full square cannot have side length 3, 4 is the smallest length possible.

Problem 2.25 Given coordinates $A = (0,4)$, $B = (4,4)$, $C = (4,0)$, $D = (1,0)$, $E = (0,0)$, and $F = (0,1)$, what is the ratio of the area of $\triangle BFD$ to the area of square $ABCE$?

Answer

$7 : 32$

Solution

Plotting the coordinates, we produce the following graph:

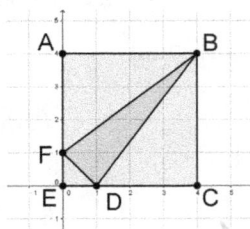

Note that the area of BFD can be determined by finding the area of the square and subtracting the areas of triangles ABF, BCD, and DEF. The area of triangles ABF and BCD is

$$\frac{1}{2} \times 3 \times 4 = 6$$

and the area of triangle DEF is

$$\frac{1}{2} \times 1 \times 1 = \frac{1}{2}.$$

Thus area of the triangle BFD is

$$16 - 2 \times 6 - \frac{1}{2} = \frac{7}{2}$$

and the ratio of the area of $\triangle BFD$ to the area of square $ABCE$ is

$$\frac{7}{2} : 16 = 7 : 32.$$

3. A Dance with Angles

Problem 3.1 Classify each of the following angles and give an estimate of the angle in degrees.

(a)

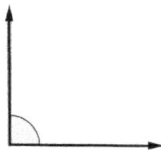

Answer

right: 90°

(b)

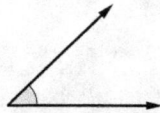

Answer

acute: 45°

(c)

Answer

straight: 180°

(d)

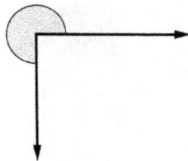

Answer

reflex: 270°

(e)

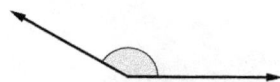

Answer

obtuse: $150°$

(f)

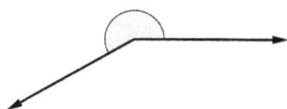

reflex: $210°$

Problem 3.2 Let m and n be a pair of parallel lines and let transversal l cut across the parallel lines as shown in the figure below, which is used in the next four questions.

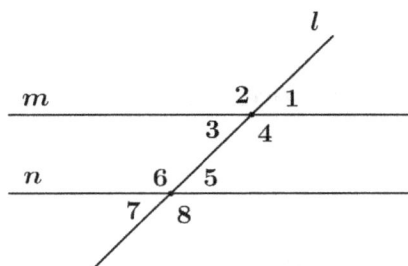

List all corresponding angles in the figure above. List all vertical angles in the figure above.

Corresponding angles: $\angle 1 = \angle 5$; $\angle 2 = \angle 6$; $\angle 3 = \angle 7$; $\angle 4 = \angle 8$ Vertical angles: $\angle 1 = \angle 3$; $\angle 2 = \angle 4$; $\angle 5 = \angle 7$; $\angle 6 = \angle 8$

Problem 3.3 Let's look at alternating angles!

(a) List all alternating interior angles in the figure above.

Answer

$$\angle 3 = \angle 5; \angle 4 = \angle 6$$

(b) **List all alternating exterior angles in the figure above.**

Answer

$$\angle 1 = \angle 7; \angle 2 = \angle 8$$

Problem 3.4 Let's look at same-side angles!

(a) **List all same-side interior angles in the figure above.**

Answer

$$\angle 4 + \angle 5 = 180°; \angle 3 + \angle 6 = 180°$$

(b) **List all same-side exterior angles in the figure above.**

Answer

$$\angle 1 + \angle 8 = 180°; \angle 2 + \angle 7 = 180°$$

Problem 3.5 In the diagram above, if $\angle 1 = 42°$, find the measures of the other angles.

Answer

$$\angle 1 = \angle 3 = \angle 5 = \angle 7 = 42° \text{ and } \angle 2 = \angle 4 = \angle 6 = \angle 8 = 138°$$

Problem 3.6 Consider the diagram below:

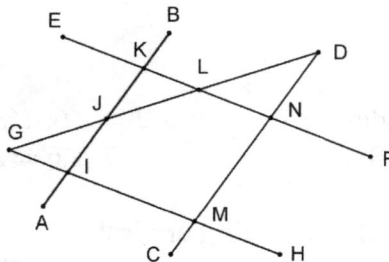

Suppose we know that $\overleftrightarrow{AB}$ and $\overleftrightarrow{CD}$ are parallel, $\angle DMH = 70°$, $\angle ELD = 135°$, and $\angle AJG = 30°$. What is $\angle LND$?

Answer

   $105°$

Solution

We first have using adjacent angles

$$\angle DLN = 180° - 135° = 45°.$$

Since $\overline{AB} \parallel \overline{CD}$,

$$\angle NDL = \angle AJG = 30°$$

as they are corresponding angles. Therefore, as the angles in triangle $\triangle DNL$ add up to $180°$,

$$\angle LND = 180 - 45 - 30 = 105°.$$

Problem 3.7 Consider the following "star" diagram, not drawn to scale.

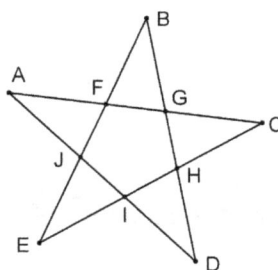

Suppose $\angle EFC = 115°$, $\angle AIC = 95°$, and $\angle BEC = 30°$. Calculate $\angle CAD$.

Answer

   $60°$.

Solution

From adjacent angles we know

$$\angle AFJ = 180° - 115° = 65°$$

and

$$\angle AIE = 180° - 115° = 85°.$$

Using $\triangle JEI$ we calculate

$$\angle EJI = 180° - 30° - 85° = 65°.$$

Hence using vertical angles,

$$\angle AJF = \angle EJI = 65°.$$

Thus

$$\angle CAD = \angle JAF = 180° - 65° - 65° = 50°.$$

Problem 3.8 **Suppose that in the diagram below we have parallel lines and a transversal.**

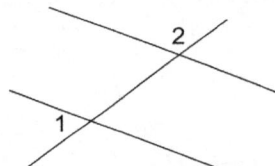

If the measure of $\angle 1$ is one-third the measure of $\angle 2$, find the measure of $\angle 1$.

Answer

45°.

Solution

Angles $\angle 1, \angle 2$ are same-side exterior angles, hence they are supplementary.
As

$$\angle 2 = 3 \times \angle 1,$$

we have

$$\angle 1 + \angle 2 = 4 \times \angle 1 = 180°$$

and hence

$$\angle 1 = 180° \div 4 = 45°.$$

Problem 3.9 Consider the diagram below:

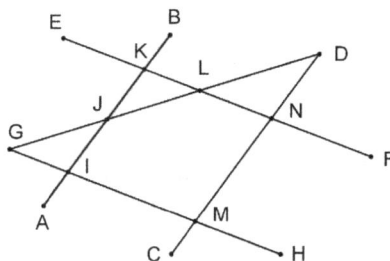

Suppose we know that $\overleftrightarrow{AB}$ and $\overleftrightarrow{CD}$ are parallel, $\angle DMH = 75°$, $\angle ELD = 140°$, and $\angle AJG = 35°$. What is $\angle BKE$?

Answer

105°

Solution

We first have

$$\angle KLJ = 180° - \angle ELD = 180° - 140° = 40°$$

by supplementary angles. Furthermore, we have

$$\angle LJK = \angle AJG = 35°$$

by vertical angles. Therefore,

$$\angle LKJ = 180° - \angle LJK - \angle KLJ = 180° - 35° - 40° = 105°.$$

Lastly, by vertical angles,

$$\angle BKE = \angle LKJ = 105°.$$

Problem 3.10 Consider the following "star" diagram, not drawn to scale.

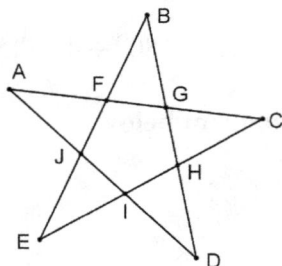

Suppose $\angle EFC = 120°$, $\angle AIC = 100°$, and $\angle BEC = 35°$. Calculate $\angle FJI$.

Answer

115°.

Solution

From adjacent angles we know

$$\angle AFJ = 180° - 120° = 60°$$

and similarly

$$\angle AIE = 180° - 100° = 80°.$$

Using $\triangle JEI$ we calculate

$$\angle EJI = 180° - 35° - 80° = 65°.$$

Hence

$$\angle FJI = 180° - 65° = 115°$$

using adjacent angles.

Problem 3.11 (2000 AMC 8) If $\angle A = 20°$ and $\angle AFG = \angle AGF$, then what is $\angle B + \angle D$?

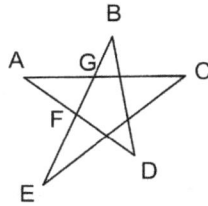

Answer

80°

Solution

We know that
$$\angle A + \angle AFG + \angle AGF = 180°$$
in a triangle, so since $\angle AFG = \angle AGF$,
$$\angle AFG = (180° - 20°) \div 2 = 80°.$$

The supplement of this angle is $180° - 80° = 100° = \angle BFD$. If we consider $\triangle BFD$, we have
$$\angle B + \angle D + 100° = 180°$$
or
$$\angle B + \angle D = 180° - 100° = 80°.$$

Problem 3.12 Given the figure below, if $\angle DFE = 75°$ and $\angle BCF = 95°$, what is the measure of $\angle CAF$?

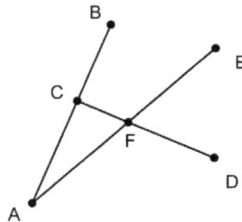

Answer

20°

Solution

Given $\angle DFE = 75°$, we observe that by vertical angles, $\angle AFC = 75°$.
Given $\angle BCF = 95°$, we get

$$\angle FCA = 180° - 95° = 85°$$

using supplementary angles. Therefore,

$$\angle AFC = 180° - 75° - 85° = 20°.$$

Problem 3.13 Suppose $\angle AOF = 180°$ and is divided into five equal angles as shown below.

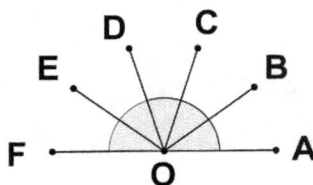

If $\angle AOB = \angle BOC = \angle COD = \angle DOE = \angle EOF$, find $\angle AOE$.

Answer

144°

Solution

All five angles are equal so each angle is

$$180 \div 5 = 36°.$$

Since $\angle AOE$ contains four of the smaller angles,

$$\angle AOE = 4 \times 36° = 144°.$$

Problem 3.14 Using the above diagram, what is $\angle BOD$?

Answer

72°

Solution

All five angles are equal so each angle is

$$180 \div 5 = 36°.$$

Since $\angle AOE$ contains two of the smaller angles,

$$\angle AOE = 2 \times 36° = 72°.$$

Problem 3.15 Consider the diagram below, where l and m are parallel but the drawing is not necessarily to scale.

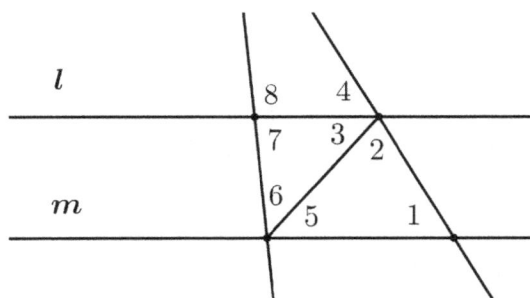

Suppose that $\angle 1 = 60°, \angle 5 = 50°, \angle 8 = 105°$. Find the measure of $\angle 6$.

Answer

55°

Solution

Since l and m are parallel, angle 8 and the angle made up of angles 5 and 6 put together are corresponding angles. Hence

$$105° = \angle 8 = \angle 6 + \angle 5 = \angle 6 + 50°.$$

Therefore,

$$\angle 6 = 105° - 50° = 55°.$$

Problem 3.16 **Using the above diagram with the given measurements, find the measure of $\angle 3$.**

Answer

$50°$

Solution

Given that lines l and m are parallel, we have that angle 1 and the angle made up of angles 2 and 3 are same-side interior angles. Therefore,

$$\angle 1 + \angle 2 + \angle 3 = 180°.$$

Furthermore, note that

$$\angle 1 + \angle 2 = 180° - \angle 5$$

as angles in a triangle add up to $180°$. Therefore,

$$180° = \angle 1 + \angle 2 + \angle 3 = 180° - \angle 5 + \angle 3,$$

so after simplifying we have $\angle 3 = \angle 5$. We know $\angle 5 = 50°$, so $\angle 3 = 30°$ as well.

Problem 3.17 **Suppose you have $\triangle ABC$ with angles $\angle A, \angle B, \angle C$. If $\angle A$ and $\angle C$ are complementary and $\angle B$ is three times $\angle A$, what is $\angle C$?**

Answer

$60°$

Solution

Angles A, C are complementary, so they add up to $90°$. Since all three angles must add up to $180°$ in a triangle, we see that

$$\angle B = 180° - 90° = 90°.$$

Therefore

$$\angle A = 90 \div 3 = 30°.$$

Lastly,

$$\angle C = 90 - 30 = 60°.$$

Problem 3.18 Consider the diagram below, where l and m are parallel but the drawing is not necessarily to scale.

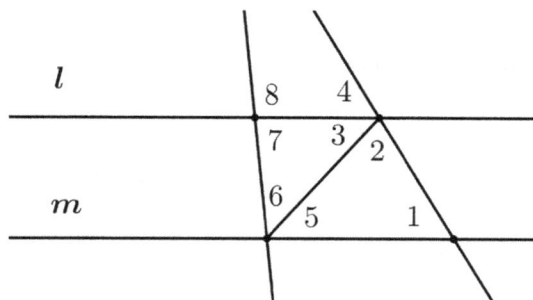

Suppose that $\angle 1 = 60°, \angle 5 = 50°, \angle 8 = 105°$. Find the measure of $\angle 4$.

Answer

   $60°$

Solution

   Since l and m are parallel, by corresponding angles, $\angle 1 = \angle 4$. Therefore, as $\angle 1 = 60°$, $\angle 4 = 60°$.

Problem 3.19 Using the above diagram with the given measurements, find the measure of $\angle 2$.

Answer

   $70°$

Solution

   We know that $\angle 1 = 60°$ and $\angle 5 = 50°$. Angles in a triangle add up to $180°$, so we have

   $$\angle 2 = 180° - 60° - 50° = 70°.$$

Problem 3.20 **(2014 AMC 8)** In $\triangle ABC$, D is a point on side $\overline{AC}$ such that $\angle BCD = \angle CBD = 70°$. **What is the degree measure of $\angle ADB$?**

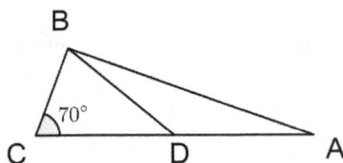

Answer

    $140°$

Solution

    Note that $\angle DBC = \angle DCB = 70°$ so

$$\angle CDB = 180° - 70° - 70° = 40°.$$

    The supplementary angle of $\angle CDB$ is

$$180° - 40° = 140°.$$

Problem 3.21 **Suppose we have the following diagram for the next three problems.**

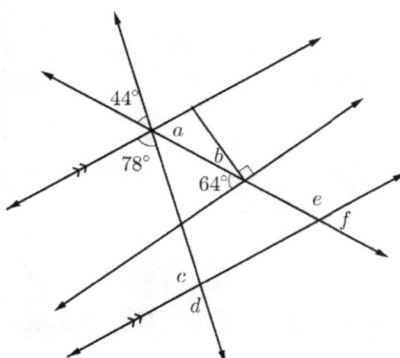

Find angles a and b.

Answer

$a = 58°, b = 26°$

Solution

Using vertical angles,

$$44° + a + 78° = 180°,$$

so

$$a = 180 - 78 - 44 = 58°.$$

Similarly, angle b is complementary to $64°$, so

$$b = 90 - 64 = 26°$$

is the measure of b.

Problem 3.22 **Find angles c and d.**

Answer

$c = 102°, d = 78°$

Solution

Firstly, we already know the missing angle vertical to a has measure $58°$. Therefore, angle c has measure

$$44 + 58 = 102°$$

since the angles are corresponding. Then as c and d are supplementary,

$$d = 180 - 102 = 78°.$$

Problem 3.23 **Find angles e and f.**

Answer

$e = 122°, f = 58°$

Solution

Angle f corresponds to angle a so angle $f = 58°$. Thus using supplementary angles,
$$e = 180 - 58 = 122°.$$

Problem 3.24 **The measures of the three angles of a triangle form an arithmetic sequence, If the smallest angle has measure $33°$, what is the angle measure of the largest angle?**

Answer

87°

Solution

The three angles of the triangle form an arithmetic sequence, so the angles have measures
$$33°, (33 + x)°, (33 + 2x)°$$
for some value of x. The three angles must add up to $180°$ so
$$33 + 33 + x + 33 + 2x = 180$$
and combining like terms gives
$$3x = 180 - 99 = 81$$
so $x = 27$. The largest angle will have measure $33 + 2 \times 27 = 87$ degrees.

Problem 3.25 **(MATHCOUNTS 2003) Two complementary angles, A and B, have measures in the ratio of 7 to 23, respectively. What is the ratio of the measure of the complement of angle A to the measure of the complement of angle B? Express your answer as a common fraction.**

Answer

$$\frac{23}{7}$$

Solution

We know that

$$\angle A : \angle B = 7 : 23.$$

A and B are complementary, so

$$\angle A + \angle B = 90°$$

so we know that

$$\angle A = 90° - \angle B, \angle B = 90° - \angle A.$$

Thus, the complements $90° - \angle A, 90° - \angle B$ are in ratio

$$90° - \angle A : 90° - \angle B = \angle B : \angle A = 23 : 7.$$

Writing as a fraction we have the ratio is $\frac{23}{7}$.

4. Magic in Triangles

Problem 4.1   (a) **Draw 2 different equilateral triangles.**

Answer

Answers may vary. Draw your own! Here is one example:

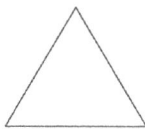

(b) **Draw 2 different isosceles triangles.**

Answer

Answers may vary. Draw your own! Here is one example:

(c) **Draw 2 different scalene triangles.**

Answer

> Answers may vary. Draw your own! Here is one example:

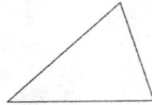

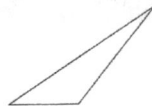

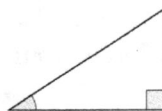

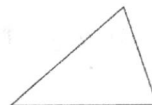

(d) **Draw 2 different obtuse triangles.**

Answer

> Answers may vary. Draw your own! Here is one example:

(e) **Draw 2 different right triangles.**

Answer

> Answers may vary. Draw your own! Here is one example:

(f) **Draw 2 different acute triangles.**

Answer

> Answers may vary. Draw your own! Here is one example:

Problem 4.2 State whether you think each of the triangles below is acute, obtuse, or right.

(a)

Answer

acute

(b)

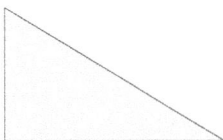

Answer

right

(c)

Answer

obtuse

(d)

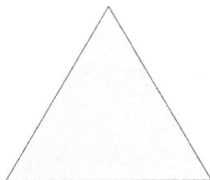

Answer

acute

(e)

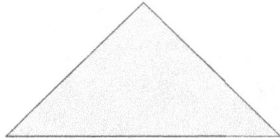

Answer

right

(f)

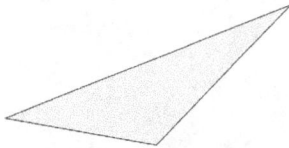

Answer

obtuse

Problem 4.3 Consider the diagram below:

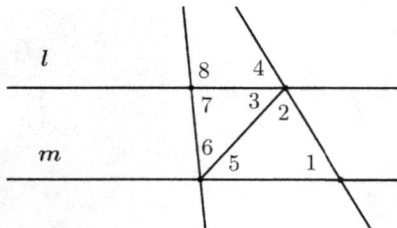

Suppose $\angle 1 = 45°, \angle 2 = 75°, \angle 6 = 60°, \angle 7 = 70°$. **Are l and m parallel?**

Answer

No

Solution

Note $\angle 3, \angle 5$ are adjacent interior angles. We have

$$\angle 5 = 180° - \angle 1 - \angle 2 = 180° - 45° - 75° = 60°$$

while

$$\angle 3 = 180° - \angle 6 - \angle 7 = 180° - 60° - 70° = 50°.$$

Since

$$\angle 3 \neq \angle 5,$$

lines l and m are not parallel.

Problem 4.4 Prove that alternate interior angles are equal and that same-side exterior angles are supplementary using the earlier results proved in class. That is, prove (for example) that $\angle 3 = \angle 5$ and that $\angle 1 + \angle 8 = 180°$ in the diagram below. Try to only use facts about vertical angles, adjacent angles, and corresponding angles in your proof.

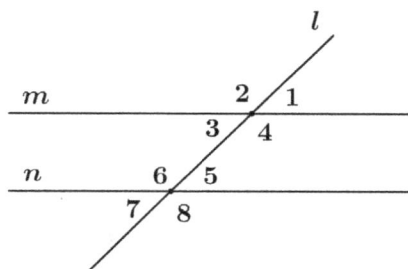

Solution

First note that

$$\angle 1 = \angle 5$$

because they are corresponding angles. Then

$$\angle 1 = \angle 3$$

using vertical angles, so

$$\angle 3 = \angle 5$$

for alternating interior angles as needed.

For the same-side exterior angles, note

$$\angle 5 + \angle 8 = 180°$$

as they are adjacent angles. Substituting ($\angle 1 = \angle 5$) we get

$$\angle 1 + \angle 8 = 180°$$

as needed.

Problem 4.5 **Suppose a triangle has a $60°$ angle and the two sides adjacent to it are equal. Is it an equilateral triangle?**

Answer

Yes

Solution

If the two sides adjacent to it are equal, then it must be an isosceles triangle. Hence, the other two angles must be equal and add up to

$$180 - 60 = 120°.$$

Therefore each angle is
$$120° \div 2 = 60°.$$

Since all three angles are $60°$, the triangle is an equilateral triangle.

Problem 4.6 **Prove that in a triangle an exterior angle is equal to the sum of the two interior angles not adjacent to it.**

Solution

The exterior angle and the interior angle adjacent to it sum to $180°$. We also know all three interior angles sum to $180°$, and the result follows.

Problem 4.7 **Prove that two angles are equal in a triangle if and only if the opposite sides are equal. (Recall that to prove an "if and only if" statement you need to prove both directions!)**

Solution

Call the triangle $\triangle ABC$. We want to show that $\angle A = \angle B$ if and only if $AC = BC$. Draw an angle bisector $\overline{CD}$, creating triangles $\triangle CAD$ and $\triangle CBD$ with $\angle BCD = \angle ACD$.

If
$$\angle A = \angle B,$$
then
$$\triangle CAD \cong \triangle CBD$$
using AAS, so
$$AC = BC,$$
proving one direction.

Similarly, if
$$AC = BC,$$
then
$$\triangle CAD \cong \triangle CBD$$
using SAS, so
$$\angle A = \angle B,$$
proving the other direction as needed.

Problem 4.8 Suppose an isosceles triangle has an angle of $30°$. Find all possibilities for the remaining two angles.

Answer

$75°, 75°$ or $30°, 120°$

Solution

It is possible for $30°$ to be either the angles repeated twice or the third angle with the other two angles repeated. If there are two $30°$ angles, the remaining angle is
$$180° - 30° - 30° = 120°.$$
Otherwise two equal angles must add up to
$$180° - 30° = 150°,$$

so are each equal to
$$150° \div 2 = 75°.$$

Problem 4.9 **Consider the diagram below.**

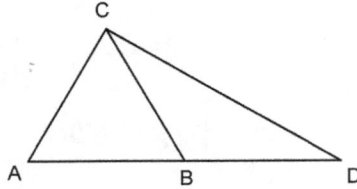

Suppose $AB = AC = 1$ **and** $\angle BAC = 60°$ **and** $\angle ADC = 30°$. **Find** BD.

Answer

1

Solution

Note using SAS, $\triangle ABC$ is an equilateral triangle. As
$$\angle CBD = 180° - 60° = 120°,$$
we have
$$\angle BCD = 180° - 120° - 30° = 30°,$$
so $\triangle BCD$ is an isosceles triangle. Hence
$$BD = BC = AB = 1.$$

Problem 4.10 **Prove that if you connect the midpoints of the sides of a triangle it divides the triangle into four smaller congruent triangles.**

Solution

Call the triangle $\triangle ABC$ with midpoints D, E, F of $\overline{AB}, \overline{BC}, \overline{CA}$. As in an earlier example, as D, E are midpoints,
$$AB = 2 \times BD$$

and
$$BC = 2 \times BE$$
so using SAS we know that $\triangle ABC \sim DBE$ with a ratio of sides $2 : 1$. A similar argument gives
$$\triangle ABC \sim \triangle ADF$$
and
$$\triangle ABC \sim \triangle FEC$$
each with ratio of sides $2 : 1$.

Therefore,
$$\triangle DBE \cong \triangle ADF \cong \triangle FEC.$$

Finally, as $\triangle DEF$ shares sides with these three triangles, using SSS we have that $\triangle DEF$ is congruent to all the other triangles as needed.

Problem 4.11 **Prove that the diagonals of a parallelogram bisect each other.**

Solution

Label the vertices of the parallelogram $ABCD$ and the intersection of the diagonals E as shown in the following diagram.

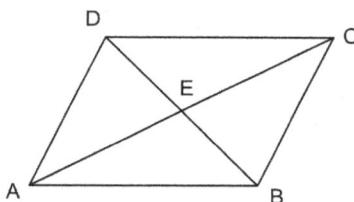

We already know from an earlier question that $\triangle ABD \cong \triangle CBD$ so
$$AB = CD.$$

By alternate interior angles, we know
$$\angle BAC = \angle DCA$$
and
$$\angle ABD = \angle CDB.$$

Thus, using ASA,
$$\triangle ABE \cong \triangle CDE.$$

Therefore $BE = DE$ and $AE = CE$ so the diagonals bisect each other.

Problem 4.12 **Prove that the perpendicular bisectors of the three sides of a triangle $\triangle ABC$ all meet in a single point.**

Solution

Proceed very similarly to the example problem involving perpendicular bisectors. Let D, E be midpoints of $\overline{AB}$ and $\overline{BC}$ in $\triangle ABC$. Let F be the intersection of the perpendicular bisectors of $\overline{AB}$ and $\overline{BC}$. As in the example, using SAS we have

$$\triangle ADF \cong \triangle BDF$$

and similarly

$$\triangle BEF \cong CEF.$$

Therefore we have $AF = BF$ and $BF = CF$ so

$$AF = BF = CF.$$

Now let G be on $\overline{AC}$ so that $\overline{AC} \perp \overline{GF}$. Note we are not assuming G is the midpoint of $\overline{AC}$, just that it is perpendicular. However, note that, using SAS again,

$$\triangle AFG \cong \triangle CFG,$$

so $AG = CG$ and G is the midpoint as needed.

Problem 4.13 **For each of the following, suppose you have $\triangle ABC$ satisfying the information given. Calculate as much of the missing information as you can. It may help to draw a diagram.**

(a) $a = b = c = 2$.

Answer

$$\angle A = \angle B = \angle C = 60°$$

Solution

As all sides are equal, the triangle is equilateral, hence

$$\angle A = \angle B = \angle C = 60°.$$

(b) $a = b = 4$ **and** $\angle C = 40°$.

Answer

$$\angle A = \angle B = 70°$$

Solution

As $a = b$, the triangle is isosceles and

$$\angle A = \angle B = (180° - 40°) \div 2 = 70°.$$

In fact it is also true that c can only be one value, but it is beyond the scope of this book to calculate it now.

(c) $\angle A = 50°, \angle C = 80°, a = 10$.

Answer

$$\angle B = 50°, b = 10$$

Solution

First we have

$$\angle B = 180° - \angle A - \angle C = 180° - 50° - 80° = 50°.$$

So $\triangle ABC$ is isosceles and hence $b = a$ so

$$b = 10.$$

As above, c can only be one value, but it is beyond the scope of this book to calculate it now.

(d) $\angle A = 40°, \angle B = 70°, a = 10$.

Answer

$$\angle C = 70°, c = 10$$

Solution

Similar to above

$$\angle C = 180° - \angle A - \angle B = 180° - 40° - 70° = 70°.$$

So $\triangle ABC$ is isosceles and hence

$$c = 10.$$

Again, b can only be one value, but it is beyond the scope of this book to calculate it now.

Problem 4.14 **Suppose the two heights outside an obtuse triangle are the same length. Prove that the triangle is isosceles.**

Solution

Let the triangle be $\triangle ABC$ with heights $BD = CE$ as shown.

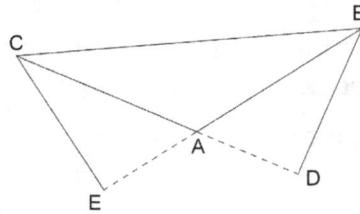

We have $BD = CE$ as a given. Both

$$\angle BDA, \angle CEA$$

are right angles and hence equal. Lastly,

$$\angle CAE = \angle BAD$$

as they are vertical angles. Therefore, by AAS,

$$\triangle CAE \cong \triangle BAD$$

and hence $CA = BA$ as needed.

Problem 4.15 **Suppose we have a "star" diagram as below (do not assume it is drawn to scale).**

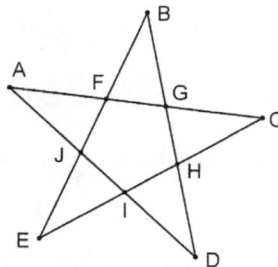

Now suppose that $\angle A = 30°$ **and that** $\triangle AFJ$ **is isosceles. Calculate** $\angle B + \angle D$.

Answer

    $75°$

Solution

We know $\triangle AFJ$ is isosceles, so

$$\angle AFJ = \angle AJF = (180° - 30°) \div 2 = 75°.$$

Hence using adjacent angles

$$\angle FJI = 180° - 75° = 105°.$$

Noting that the angles in $\triangle BJD$ must add up to $180°$, we have

$$\angle B + \angle D = 180 - 105 = 75°.$$

Problem 4.16 (2006 AMC 8) Triangle ABC is an isosceles triangle with $\overline{AB} = \overline{BC}$. **Point** D **is the midpoint of both** $\overline{BC}$ **and** $\overline{AE}$, **and** $\overline{CE}$ **is** 11 **units long. Triangle** ABD **is congruent to triangle** ECD.

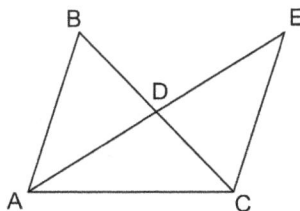

What is the length of $\overline{BD}$?

Answer

    5.5

Solution

Since triangle ABD is congruent to triangle ECD and $\overline{CE} = 11$, it follows that

$$\overline{AB} = 11.$$

Furthermore, since $\overline{AB} = \overline{BC}$, $\overline{BC} = 11$ as well. Because point D is the midpoint of $\overline{BC}$,

$$BD = BC \div 2 = 11 \div 2 = 5.5.$$

Problem 4.17 Suppose that $ABCD$ is a square. Let point E be *outside* the square and that $\triangle CDE$ is an equilateral triangle (see the diagram). What is the measure of $\angle EAD$?

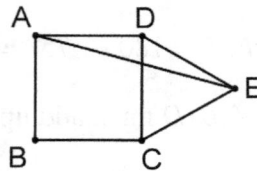

Answer

$15°$

Solution

All the sides of a square are equal as are sides of an equilateral triangle, so $DE = DC = DA$, and thus $\triangle ADE$ is isosceles and $\angle EAD = \angle DEA$. We know that

$$\angle ADE = 90° + 60° = 150°,$$

and

$$\angle ADE + \angle EAD + \angle DEA = 180°,$$

thus

$$\angle EAD = (180° - 150°) \div 2 = 15°.$$

Problem 4.18 Given square $ABCD$, let P and Q be the points outside the square that make triangles CDP and BCQ equilateral. Segments $\overline{AQ}$ and $\overline{BP}$ intersect at G. Find angle AGP.

Answer

90°

Solution

An identical argument to previous the problems gives

$$\angle CBP = \angle BAQ = 15°.$$

Hence

$$\angle ABG = 90° - 15° = 75°,$$

and then

$$\angle BGA = 180° - 75° - 15° = 90°$$

so

$$\angle AGP = 180° - 90° = 90°$$

as well.

Problem 4.19 **Given square** $ABCD$, **let** P **and** Q **be the points outside the square that make triangles** CDP **and** BCQ **equilateral. Prove that triangle** APQ **is also equilateral.**

Solution

To prove $AP = PQ = QA$, we prove

$$\triangle ADP \cong \triangle QBA \cong \triangle QCP$$

via SAS congruency: First we have

$$AB = BQ = QC = CP = DP = AD$$

as all are either sides of the square or the equilateral triangles. Secondly, we have

$$\angle ADP = \angle QBA = \angle QCP = 150°.$$

Hence we can use SAS congruency as claimed.

Problem 4.20 **Suppose in the diagram below that** $\triangle ABC$ **is isosceles and** $\angle CAG = 20°$. **Find the measure of** $\angle ABD$.

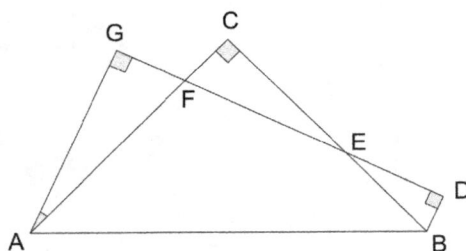

Answer

115°

Solution

First note as $\triangle ABC$ is an isosceles right triangle,

$$\angle ABC = (180° - 90°) \div 2 = 45°.$$

Since $\angle CAG = 20°$,

$$\angle AFG = 90° - 20° = 70°.$$

Continuing in this manner, we see

$$\angle CAG = \angle CEG = \angle DEB = 20°$$

and

$$\angle AFG = \angle CFE = \angle DBE = 70°.$$

Therefore,

$$\angle ABD = 70° + 45° = 115°.$$

Problem 4.21 (2015 AMC 8) What is the smallest whole number larger than the perimeter of any triangle with a side of length 5 and a side of length 19?

Answer

48

Solution

Let s be the last side of the triangle. We want a whole number larger than any possible perimeter, so we try to make s as large as possible, so assume that $s > 19$.

Using the triangle inequality, we know that

$$s < 5 + 19 = 24.$$

Therefore, the perimeter P is

$$P = s + 5 + 19 < 24 + 5 + 19 = 48$$

so the perimeter P is less than 48.

Note that a triangle with sides $5, 19, 23$ (with perimeter 47) is possible, so 48 is our answer.

Problem 4.22 **Suppose triangle ABC is formed by the coordinates $A = (0,0)$, $B = (2,0)$, and $C = (0,4)$. A second congruent triangle DEF is formed with the two given points $D = (0,0)$ and $E = (0,2)$. If at least one of the coordinates of F is negative, what are the possible coordinates of point F?**

Answer

$(-4, 0)$ or $(-4, 2)$

Solution

Note that AB has length 2 and AC has length 4. Further, we know that

$$\angle BAC = 90°.$$

We know that $DE = 2$ so it corresponds to side AB. The two possibilties are

$$\angle DEF = 90°$$

or

$$\angle EDF = 90°.$$

Hence the y-coordinate of F is either 0 or 2. Thus the x-coordinate must be negative. If $\angle DEF = 90°$, then $F = (-4, 2)$ and if $\angle EDF = 90°$ then $F = (-4, 0)$.

Problem 4.23 Suppose that $ABCD$ is a square. Let point E be *inside* the square and that $\triangle CDE$ is an equilateral triangle. What is the measure of $\angle EAD$?

Answer

$75°$

Solution

We have
$$\angle AED = 90° - 60° = 30°.$$

As the sides of the square and equilateral triangle are all equal, $\triangle AED$ is isosceles. We therefore have
$$\angle EAD = (180° - 30°) \div 2 = 75°.$$

Problem 4.24 (ZIML 2016) In $\triangle ABC$, $AB = AC$. Point D is on side $\overline{AB}$ such that $\overline{CD}$ bisects $\angle ACB$, and $CD = BC$. Find the measure of $\angle BAC$ in degrees.

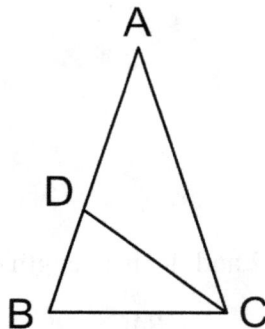

Answer

$36°$

Solution

Let $\angle ABC$ have measure x. Given that $AB = AC$, triangle ABC is isosceles with $\angle ACB = x$. Given that CD bisects $\angle ACB$, so

$$\angle BCD = \angle ACD = \frac{x}{2}.$$

Since $CD = BC$, we also have

$$\angle CDB = \angle DBC = x.$$

Therefore,

$$x + x + \frac{x}{2} = 180°$$

so

$$\frac{5}{2}x = 180°$$

and then $x = 72°$. It follows that

$$\angle BAC = 180° - 72° - 72° = 36°.$$

Problem 4.25 (2005 AMC 8) How many different isosceles triangles have integer side lengths and perimeter 23?

Answer

6

Solution

Start off by defining the sides of the triangle to be a, a, and b. Therefore we want the perimeter

$$2a + b = 23$$

so $b = 23 - 2a$. Since the side lengths are positive, the maximum value of a is 11, leading to a value of $b = 1$. Furthermore, by the Triangle Inequality, we see that

$$2a > b = 23 - 2a$$

which simplifies to

$$4a > 23.$$

Therefore, a can have integer side lengths from 6 to 11 so there are a total of 6 triangles. You can calculate these side lengths to be:

$$(6,6,11), (7,7,9), (8,8,7), (9,9,5), (10,10,3), (11,11,1).$$

5. You Are Special, Right?

Problem 5.1 **If one of the leg lengths of a right triangle is 20 and the other leg length is 21, what is the length of the hypotenuse?**

Answer

  29

Solution

  Using the Pythagorean theorem, the hypotenuse is $\sqrt{20^2 + 21^2} = 29$.

Problem 5.2 **If one of the leg lengths of a 45-45-90 triangle is 3, what is the length of the hypotenuse?**

Answer

  $3\sqrt{2}$

Solution

Since 45-45-90 triangles have side ratios $1 : 1 : \sqrt{2}$, the length of the hypotenuse is $3 \times \sqrt{2} = 3\sqrt{2}$.

Problem 5.3 **If the shorter leg length of a 30-60-90 triangle is 3, what is the length of the hypotenuse?**

Answer

6

Solution

Since 30-60-90 triangles have side ratios $1 : \sqrt{3} : 2$, the length of the hypotenuse is $2 \times 3 = 6$.

Problem 5.4 **Consider the diagram below.**

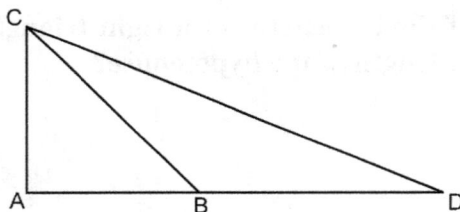

Suppose $AB = AC = 1$ and $\angle BAC = 90°$ and $\angle ADC = 22.5°$. Find BD.

Answer

$\sqrt{2}$

Solution

Note $\triangle ABC$ is an isosceles right triangle, so it is a 45-45-90 triangle. Thus $BC = \sqrt{2}$. As

$$\angle CBD = 180° - 45° = 135°,$$

we have
$$\angle BCD = 180° - 135° - 22.5° = 22.5°,$$
so $\triangle BCD$ is an isosceles triangle. Hence $BD = BC = \sqrt{2}$.

Problem 5.5 Let ABC be a right triangle with leg lengths 8 and 15. All lengths of ABC are doubled to form new triangle DEF. Find the ratio of the area of ABC to the area of DEF.

Answer

1 : 4

Solution

The area of triangle ABC is
$$\frac{1}{2} \times 8 \times 15 = 60.$$

The legs of the legs in triangle DEF are 16 and 30. Therefore, the area of triangle DEF is
$$\frac{1}{2} \times 16 \times 30 = 240.$$
Since $240 \div 60 = 4$ the ratio of areas is 1 : 4.

Problem 5.6 Given two squares, show how to cut them up into pieces that can be combined to form one larger square. Your method should work no matter which two squares you are given. It is probably easiest to start with the squares side by side as in the diagram below.

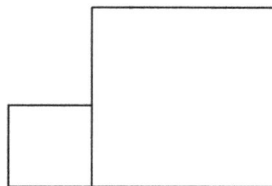

Solution

Make the cuts below as in the diagram on the left (where the two triangles cut off are congruent) and rearrange as shown on the right:

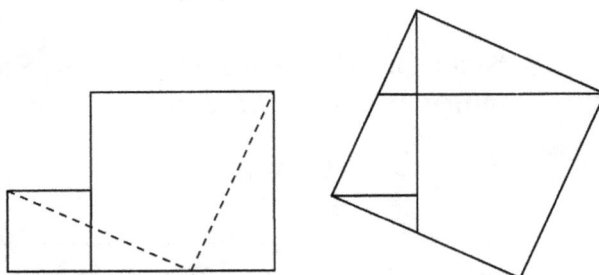

Problem 5.7 **For each of the following, state whether it is possible to have a right triangle with the given side lengths. If it is possible, we call** (a, b, c) **a** *Pythagorean Triple*.

(a) $6, 18, 21$

> **Answer**

No

(b) $7, 24, 25$

> **Answer**

Yes

(c) $8, 15, 17$

> **Answer**

Yes

(d) $9, 18, 27$

> **Answer**

No

(e) $10, 24, 26$

> **Answer**

Yes

Problem 5.8 Let $ABCD$ be a rectangle with $AB = 18, BC = 5$. Let E be the point a third of the way from A to B on $\overline{AB}$. Is $\angle CED$ a right angle?

Answer

No

Solution

Since E is a third of the way from A to B, we have

$$AE = 18 \div 3 = 6$$

and

$$BE = 2 \times 6 = 12.$$

Therefore, we can use the Pythagorean theorem on right triangles $\triangle BCE$ and $\triangle ADE$ to get that

$$CE^2 = 169, DE^2 = 61.$$

We then see

$$18^2 = 324 = DC^2 \neq DE^2 + CE^2 = 169 + 61 = 230,$$

so $\triangle CDE$ is not a right triangle and hence $\angle CED$ is not a right angle.

Problem 5.9 (1999 AMC 8) In trapezoid $ABCD$, the sides AB and CD are equal.

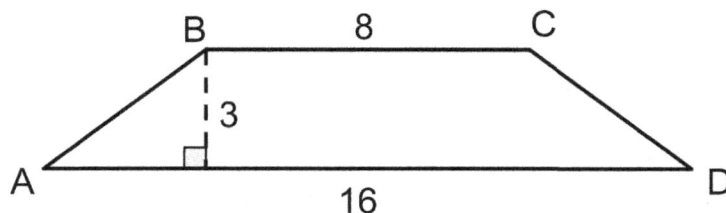

What is the perimeter of $ABCD$?

Answer

34

Solution

Cut the trapezoid into two triangles and a rectangle as in the diagram below.

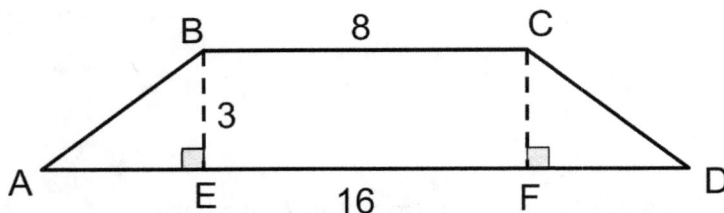

In fact, using SSS the two triangles ABE and DCF are congruent (they have the same hypotenuse and the same leg, so their third leg is also the same by the Pythagorean theorem). Hence

$$AE + EF + FD = 18$$

and since $BCEF$ is a rectangle, $EF = BC = 8$ so we get

$$AE = DF = (16 - 8) \div 2 = 4.$$

By Pythagorean theorem,

$$AB = CD = \sqrt{3^2 + 4^2} = 5.$$

The sides of the trapezoid are thus 8, 5, 16, and 5. Adding those up gives us the perimeter

$$8 + 5 + 16 + 5 = 13 + 21 = 34.$$

Problem 5.10 **Let's work with isosceles right triangles.**

(a) **The legs of an isosceles right triangle are each 4 cm long. How long is an altitude drawn to the hypotenuse of this triangle?**

Answer

$2\sqrt{2}$ cm

Solution

Drawing the altitude to the hypotenuse forms two triangles with angles $90°$ and $45°$. Hence their third angles are

$$180° - 90° - 45° = 45°$$

as well. As the triangles also share a side, they are congruent.

These triangles have hypotenuse 4 cm (the base of the original triangle). The leg of these triangles is the altitude we want, so using the fact that 45-45-90 triangles have ratio of sides $1 : 1 : \sqrt{2}$, the altitude has length

$$4 \times \frac{1}{\sqrt{2}} = \frac{4}{\sqrt{2}} = \frac{4\sqrt{2}}{2} = 2\sqrt{2}$$

cm.

(b) **The altitude of an isosceles right triangle that meets the hypotenuse of the triangle has length 3 cm. What is the length of the shortest side of the triangle?**

Answer

$3\sqrt{2}$ cm

Solution

As in the previous problem, this altitude divides the full triangle into two 45-45-90 triangles. We are given these smaller 45-45-90 triangles have legs of length 3. Hence they have a hypotenuse of

$$3 \times \frac{\sqrt{2}}{1} = 3\sqrt{2},$$

which is the leg of the full triangle as needed.

(c) **If the area of an isosceles right triangle is 2 square inches, how long is the shortest side?**

Answer

2 in

Solution

Let l be the length of the leg. Then the triangle has base l and height l, so

$$\frac{1}{2} \times l \times l = 2.$$

Thus $l^2 = 4$ so we see the shortest side has length 2 inches.

Problem 5.11 In the diagram, $\triangle ABC, \triangle DEF$ are two congruent isosceles right triangles. Given that $ADFC$ is a 4×2 rectangle, find the area of the shaded region.

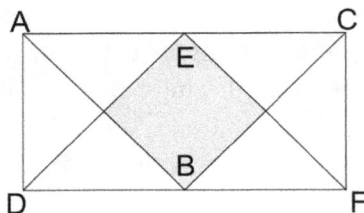

Answer

2

Solution

Note by connecting E and B, we divide $ACFD$ into 8 congruent isosceles right triangles. Therefore, the shaded region (which is 2 of these triangles) has area

$$8 \div 4 = 2.$$

Problem 5.12 Mark P inside square $ABCD$, so that triangle ABP is equilateral. Let Q be the intersection of BP with diagonal AC. Triangle CPQ looks isosceles. Is this actually true?

Answer

Yes

Solution

As in a previous problem, first note $BC = AB = BP$ so $\triangle BPC$ is isosceles. Since

$$\angle PBC = 90° - 60° = 30°,$$

we have that

$$\angle BPC = \angle BCP = (180° - 30°) \div 2 = 75°.$$

We know that $\angle ACB = 90° \div 2 = 45°$. So

$$\angle ACP = \angle BCP - \angle ACB = 75° - 45° = 30°.$$

Looking now at triangle $\triangle CPQ$, we know $\angle PCQ = 30°$ and $\angle QPC = \angle BPC = 75°$. Hence

$$\angle PQC = 180° - 30° - 75° = 75°,$$

so the triangle is isosceles.

Problem 5.13 **Given a parallelogram $ABCD$, let P and Q be the points outside the parallelogram so that triangles CDP and BCQ are equilateral. Is the triangle APQ is equilateral?**

Answer

Yes

Solution

First we have
$$AB = CD = CP = DP$$

as opposite sides of a parallelogram are equal and the equilateral triangle CDP has three equal sides. An identical argument gives

$$AD = BC = BQ = CQ.$$

Recalling that in a parallelogram the diagonal divides the parallelogram into two congruent triangles, we have that the opposite angles $\angle ABC, \angle CDA$ are equal. Thus

$$\angle ABQ = \angle ABC + 60° = \angle CDA + 60° = \angle PDA.$$

Further, since $\angle ABC, \angle CBD$ are same-side interior angles,

$$\angle CBD = 180° - \angle ABC.$$

Thus,

$$\angle PCQ = 360° - 60° - \angle CBD - 60° = 240° - (180° - \angle ABC) = \angle ABC + 60°.$$

Therefore

$$\angle PCQ = \angle ABQ = \angle PDA$$

so

$$\triangle PCQ \cong \triangle ABQ \cong \triangle PDA$$

using SAS. Thus $AQ = QP = PA$ so the triangle is equilateral as needed.

Problem 5.14 (1999 AMC 8) Square $ABCD$ has sides of length 3. Segments CM and CN divide the square's area into three equal parts.

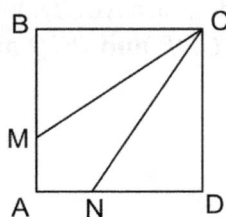

How long is segment CM?

Answer

$\sqrt{13}$

Solution

Given that the square has side length 3, the area of the square is

$$3^2 = 9.$$

Given that the segments divide the square into 3 parts containing the same area, each part has area 3. $\triangle CBM$ has area 3 and base $CB = 3$, so by using the area formula for a triangle, we get

$$3 = \frac{1}{2} \times 3BM$$

so $BM = 2$. Since $\triangle CBM$ is a right triangle,

$$CM = \sqrt{BM^2 + BC^2} = \sqrt{2^2 + 3^2} = \sqrt{13}$$

using the Pythagorean theorem.

Problem 5.15 **Let ABC be a triangle given below:**

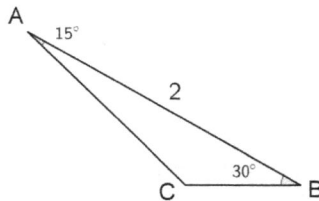

Determine the area of triangle ABC.

Answer

$$\frac{\sqrt{3} - 1}{2}$$

Solution

If we draw the altitude of the triangle we get the diagram below:

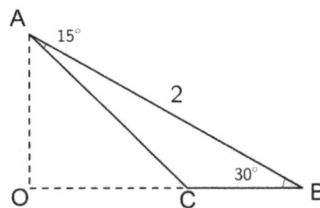

we will note that there is a 30-60-90 triangle AOB and a 45-45-90 triangle AOC present. This can be justified by noting that the obtuse angle measures

$$180° - 15° - 30° = 135°$$

so the exterior angle that is formed outside of the obtuse angle measures $45°$. Applying the side ratio relationship for 30-60-90 triangles gives us that

$$AO = 2 \times \frac{1}{2} = 1.$$

As $\triangle AOC$ a 45-45-90 triangle, we have $AO = CO = 1$. Using the 30-60-90 triangle again gives $BO = \sqrt{3}$, so

$$BC = BO - CO = \sqrt{3} - 1.$$

Hence triangle ABC has base length $\sqrt{3} - 1$ and height 1. Therefore, the area of triangle ABC is

$$\frac{1}{2} \times (\sqrt{3} - 1) \times 1 = \frac{\sqrt{3} - 1}{2}.$$

Problem 5.16 (2005 AMC 8) What is the perimeter of trapezoid $ABCD$?

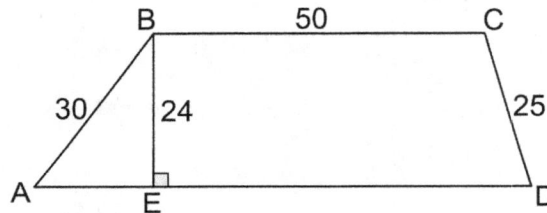

Answer

180

Solution

If we draw altitudes from B and C to base AD, we create a rectangle and two right triangles as in the diagram below.

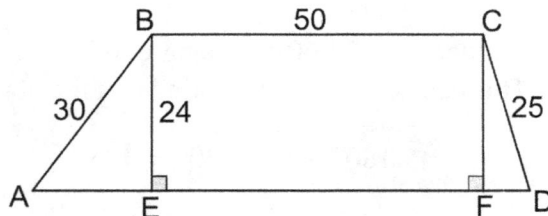

Note that the side opposite BC is equal to 50. Remember that $(3,4,5)$ is a Pythagorean triple, so

$$(6 \times 3, 6 \times 4, 6 \times 5) = (18, 24, 30)$$

is another Pythagorean triple so $AE = 18$. Similarly, $(7, 24, 25)$ is a Pythagorean triple so $DF = 7$. Add it together to get

$$AD = 18 + 50 + 7 = 75.$$

The perimeter is

$$75 + 30 + 50 + 25 = 180.$$

Problem 5.17 **Draw the largest possible square inside an equilateral triangle, with one side of the square aligned with one side of the triangle. If the equilateral triangle has side length 6, find the side length of the square.**

Answer

$12\sqrt{3} - 18$

Solution

Let x be the length of the square. Note that the square and equilateral triangle form two 30-60-90 triangles as in the diagram below

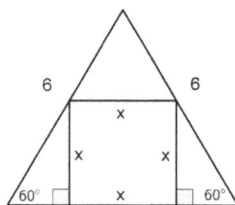

Note the bottom side of each 30-60-90 triangle is

$$x \times \frac{1}{\sqrt{3}} = \frac{x}{\sqrt{3}} = \frac{x}{3}\sqrt{3},$$

so the side length of the square satisfies

$$\frac{2x}{3}\sqrt{3} + x = 6.$$

Therefore,

$$x = \frac{18}{2\sqrt{3}+3} = \frac{18}{2\sqrt{3}+3} \times \frac{2\sqrt{3}-3}{2\sqrt{3}-3} = \frac{18(2\sqrt{3}-3)}{12-9} = 12\sqrt{3}-18.$$

Problem 5.18 (2004 AMC 8) In the figure, $ABCD$ is a rectangle and $EFGH$ is a parallelogram.

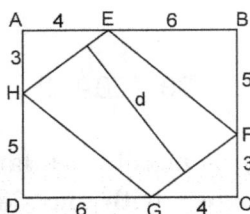

Using the measurements given in the figure, what is the length d of the segment that is perpendicular to $\overline{HE}$ and $\overline{FG}$?

Answer

7.6

Solution

Note that the area of parallelogram $EFGH$ can be determined by taking the area of the entire rectangle and subtracting area of the triangles from our determined area. Since the rectangle is a $4+6 = 10$ by $3+5 = 8$ rectangle, the area of parallelogram $EFGH$ is

$$10 \times 8 - 2 \times \frac{1}{2} \times 6 \times 5 - 2 \times \frac{1}{2} \times 3 \times 4 = 80 - 30 - 12 = 38.$$

Also note, that the area of the parallelogram $EFGH$ can also be determined using the length of the base FG and the height of the parallelogram d. By the Pythagorean theorem (recall $(3,4,5)$ is a Pythagorean triple), $FG = 5$. Therefore the area of parallelogram $EFGH$ is $5 \times d$, so we know that

$$5 \times d = 38.$$

Therefore $d = 38 \div 5 = 7.6$.

Problem 5.19 **The ratio of the hypotenuse of a right triangle to one of the other sides is** $41:9$**. What is the smallest possible area of the triangle if the length of each side of this triangle is an integer?**

Answer

180

Solution

The smallest possible hypotenuse is 41, so assume the hypotenuse is 41 so that one of the legs is 9. Using the Pythagorean theorem, the other leg is

$$\sqrt{41^2 - 9^2} = \sqrt{1681 - 81} = \sqrt{1600} = 40.$$

Thus the triangle has base 40 and height 9 and thus area

$$\frac{1}{2} \times 40 \times 9 = 180,$$

which is the smallest possible.

Problem 5.20 **A triangle has sides measuring 37 cm, 37 cm and 24 cm. A second triangle not congruent to the first, is drawn with sides 37 cm, 37 cm and x cm, where x is a whole number. If the two triangles have equal areas, what is the value of x?**

Answer

70

Solution

As in earlier problems, if we draw a height in the triangle with sides 37, 37, 24 to the 24 cm side, it divides the triangle into two congruent triangles. Note these triangles have hypotenuse 37 and one leg of

$$24 \div 2 = 12$$

cm. Hence using the Pythagorean theorem we have that the other leg (the height) has length

$$\sqrt{37^2 - 12^2} = \sqrt{1369 - 144} = \sqrt{1225} = 35.$$

Thus the first triangle has area

$$\frac{1}{2} \times 24 \times 35 = 420$$

cm^2. Note that above we really showed that $(12, 35, 37)$ is a Pythagorean triple. Since $2 \times 35 = 70$, an identical argument to that above shows that a triangle with sides 37, 37, 70 cm will also have area

$$\frac{1}{2} \times 70 \times 12 = 420$$

cm^2 as well. Therefore $x = 70$.

Problem 5.21 (ZIML 2016) In $\triangle ABC$, $\angle B = 90°$, $AB = 8$, and $BC = 6$. Extend the line segment $\overline{BC}$ to point D so that the obtuse triangle $\triangle ACD$ is formed. If $CD = 9$, find the perimeter of $\triangle ACD$.

Answer

    36

Solution

Consider the diagram:

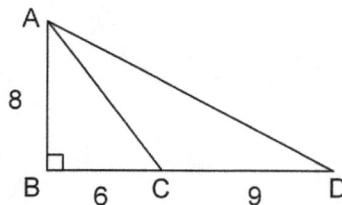

Note that when we extend BC to D, we create two right triangles ABC and ABD. Triangle ABC have leg lengths 6 and 8 and triangle ABD have leg lengths 15 and 8. Applying the Pythagorean theorem, we observe that

$$AC = \sqrt{6^2 + 8^2} = \sqrt{100} = 10$$

and

$$AD = \sqrt{15^2 + 8^2} = \sqrt{289} = 17.$$

Therefore, the perimeter of triangle ACD is

$$10 + 17 + 9 = 36.$$

Problem 5.22 **Let $A = (0,0), B = (1,1),$ and $C = (2,0)$ be coordinates. Find the angle measure of BCA?**

Answer

   $45°$

Solution

Apply the distance formula to determine the lengths of the triangle. By doing so, we have:

$$AB = \sqrt{(1-0)^2 + (1-0)^2} = \sqrt{2}$$
$$BC = \sqrt{(2-1)^2 + (0-1)^2} = \sqrt{2}$$
$$AC = \sqrt{(2-0)^2 + (0-0)^2} = 2$$

Determining the ratio of the side lengths, we observe that

$$AB : BC : AC = \sqrt{2} : \sqrt{2} : 2 = 1 : 1 : \sqrt{2}.$$

This implies that triangle ABC is a 45-45-90 triangle. Since the smaller length corresponds to the angle measure of $\angle BCA$, $\angle BCA = 45°$.

Problem 5.23 **(2015 AMC 8) In $\triangle ABC$, $AB = BC = 29$, and $AC = 42$. What is the area of $\triangle ABC$?**

Answer

   420

Solution

As in earlier problems, start by drawing an altitude to the AC, which splits the triangle into two congruent right triangles. These triangle have hypotenuse 29 and leg 21. By Pythagorean theorem, the length of the last leg has measure

$$\sqrt{29^2 - 21^2} = \sqrt{841 - 441} = \sqrt{400} = 20.$$

Therefore, the area of the triangle is

$$\frac{1}{2} \times 20 \times 42 = 420.$$

Problem 5.24 (**2012 AMC 8**) **A square with area 4 is inscribed in a square with area 5, with one vertex of the smaller square on each side of the larger square. A vertex of the smaller square divides a side of the larger square into two segments, one of length a, and the other of length b.**

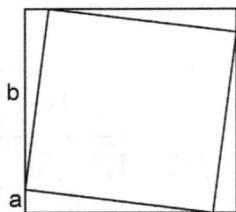

What is the value of ab?

Answer

$\frac{1}{2}$

Solution

The total area of the four congruent triangles formed by the squares is

$$5 - 4 = 1.$$

Therefore, the area of each of these triangles is

$$1 \div 4 = \frac{1}{4}.$$

Note these triangles have height a and base b. Using the formula for area of the triangle, we have that

$$\frac{1}{2} \times a \times b = \frac{1}{4}.$$

Multiply by 2 on both sides to find that the value of ab is

$$2 \times \frac{1}{4} = \frac{1}{2}.$$

Problem 5.25 **(2015 AMC 8) One-inch squares are cut from the corners of this 5 inch square.**

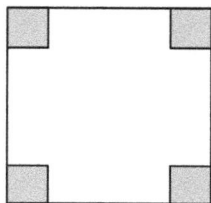

What is the area in square inches of the largest square that can be fitted into the remaining space?

Answer

15

Solution

Draw the square as shown below:

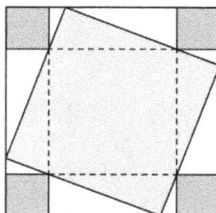

We wish to find the area of the square. Note that the area of the larger square is composed of the smaller square and the four triangles. Each triangle has base $5 - 1 - 1 = 3$ and height 1, so the combined area of the four triangles is

$$4 \times \frac{1}{2} \times 3 \times 1 = 6.$$

Since the area of the smaller square is $3^2 = 9$, the area of the large square is $9 + 6 = 15$.

6. Angles Are Special Too

Problem 6.1 **Give an example of an equiangular polygon that is not equilateral.**

Answer

Rectangle. Many examples are possible.

Problem 6.2 **Give an example of an equilateral polygon that is not equiangular.**

Answer

Rhombus. Many examples are possible.

Problem 6.3 **Inside regular pentagon** $ABCDE$ **is marked point** F **so that triangle** ABF **is equilateral. Decide whether or not quadrilateral** $ABCF$ **is a parallelogram, and give your reasons.**

Answer

No

Solution

Note that $\angle BAF = 60°$ as ABF is an equilateral triangle. $\angle ABC = 108°$ as it is the interior angle of a regular pentagon. If $\overline{AF} \parallel \overline{BC}$, then $\angle BAF$ and $\angle ABC$ would be same-side interior angles and hence supplementary. However,

$$60° + 108° = 168° \neq 180°,$$

so the lines are not parallel and thus $ABCF$ is not a parallelogram. In fact the angles of $ABCF$ can be calculated as $60°, 108°, 66°, 126°$.

Problem 6.4 A regular nonagon has nine sides. Find the interior angle sum, the exterior angle sum, and the interior angle measure of a nonagon.

Answer

Interior angle sum: $1260°$, Exterior angle sum: $360°$, Interior angle $140°$

Solution

Using the pattern from the chart in the examples, the interior angle sum of a nonagon is

$$180 \times (9 - 2) = 1260°,$$

the exterior angle sum is $360°$, and the interior angle measure of a regular nonagon is

$$1260 \div 9 = 140°.$$

Problem 6.5 A regular icosagon has twenty sides. Find the interior angle sum, the exterior angle sum, and the interior angle measure of a nonagon.

Answer

Interior angle sum: $3240°$, Exterior angle sum: $360°$, Interior angle $162°$

Solution

Using the pattern from the chart in the examples, the interior angle sum of a nonagon is

$$180 \times (20 - 2) = 3240°,$$

the exterior angle sum is $360°$, and the interior angle measure of a regular nonagon is

$$3240 \div 20 = 162°.$$

Problem 6.6 Use 6 **equilateral triangles to form a hexagon** $ABCDEF$**, then find the area of triangle** AED **in terms of the total area.**

Answer

$$\frac{1}{3}$$

Solution

Call O the center of the hexagon. Then $\triangle OED$ is a triangle with area $\frac{1}{6}[ABCEDF]$. Further note that $\triangle AEF \cong \triangle AOE$ using SAS. Therefore,

$$[AOE] = \frac{1}{2}[AOEF] = \frac{1}{2} \times \frac{2}{6}[ABCDEF] = \frac{1}{6}[ABCDEF].$$

Therefore,

$$[AEF] = \frac{1}{3}[ABCDEF].$$

Problem 6.7 **Equilateral triangles** BCP **and** CDQ **are attached to the outside of regular pentagon** $ABCDE$**. Is quadrilateral** $BPQD$ **a parallelogram? Justify your answer.**

Answer

No

Solution

First note that $\triangle BCD$ is isosceles, so since $\angle BCD = 108°$ we have

$$\angle CBD = \angle CDB = (180° - 108°) \div 2 = 36°.$$

Hence

$$\angle PBD = \angle QDB = 60° + 36° = 96°.$$

If $\overline{BP} \parallel \overline{DQ}$ then $\angle PBD$ and $\angle QDB$ would be same-side interior angles, hence supplementary. However

$$96° + 96° = 192° \neq 180°,$$

so $\overline{BP}$ and $\overline{DQ}$ are not parallel so $BPQD$ is not a parallelogram. In fact, it can be shown that it is a trapezoid with angles $96°, 96°, 84°, 84°$.

Problem 6.8 Three non-overlapping regular plane polygons all have sides of length 1. The polygons meet at a point A in such a way that the sum of the three interior angles at A is $360°$. Thus the three polygons form a new polygon P (not necessarily convex) with A as an interior point. Suppose two of the polygons are pentagons. Find the perimeter of P.

Answer

14

Solution

The last polygon has interior angles of

$$360 - 108 - 108 = 144°,$$

hence must be a decagon. Therefore the perimeter of the entire shape is the sum of the perimeter of each of the three polygons minus the 3 shared edges. Note each shared edge is counted twice. Hence the perimeter is

$$5 + 5 + 10 - 3 \times 2 = 14.$$

Problem 6.9 Let $ABCDEFGH$ be a regular octagon, and let $GHIJKL$ be a regular hexagon. Find all possible values of measure of $\angle IAH$.

Answer

$82.5°, 37.5°$

Solution

We proceed almost identically to the similar example. We have that either the hexagon is inside or outside the octagon. Note both $\triangle AFG_1$ and $\triangle AFG_2$ are isosceles. Recall that interior angles of a octagon are $135°$ and $120°$ in a hexagon. If the hexagon is inside the octagon, then

$$\angle AHI = 135° - 120° = 15°,$$

so

$$\angle AIH = \angle IAH = (180° - 15°) \div 2 = 82.5°.$$

If the hexagon is outside the octagon, we similarly have that

$$\angle AHI = 360° - 135° - 120° = 105°,$$

so

$$\angle AIH = \angle IAH = (180° - 105°) \div 2 = 37.5°.$$

Hence the two possibilties for $\angle IAH$ are $82.5°$ or $37.5°$.

Problem 6.10 **Suppose that** $DRONE$ **is a regular pentagon, and that** $DRU, ROC, ONL, NEA,$ **and** EDI **are equilateral triangles attached to the outside of the pentagon. Is** $IUCLA$ **a regular pentagon?**

Answer

Yes

Solution

Look at for example $\triangle UCR$. It is an isosceles triangle, and

$$\angle URC = 360 - 108 - 60 - 60 = 132.$$

Hence

$$\angle CUR = \angle UCR = \frac{1}{2}(180 - 132) = 24.$$

Similar results hold for $\triangle CLO, \triangle LAN, \triangle AIE, \triangle IUD$, so all these triangles are congruent. Hence pentagon $ICULA$ is equilateral. From this it is also clear that $ICULA$ is equiangular and thus regular as claimed.

Problem 6.11 **The equiangular convex hexagon** $ABCDEF$ **has** $AB = 10$, $BC = 10$, $CD = 8$, **and** $DE = 8$. **Find** $[ABCDEF]$.

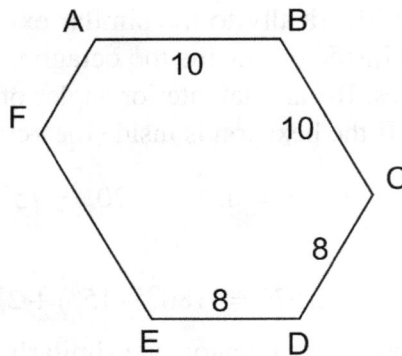

Answer

$119\sqrt{3}$

Solution

Recall that since $ABCDEF$ is an equiangular hexagon, all the interior angles are $120°$. Thus, if we extend the sides $\overline{AF}, \overline{BC}, \overline{DE}$ we form a large equilateral triangle RST as shown.

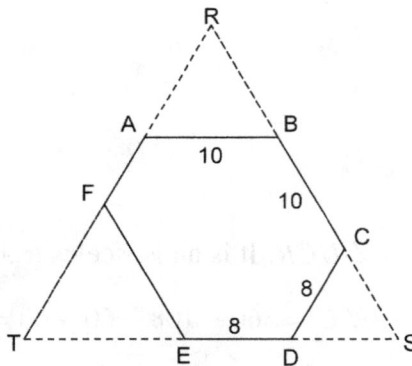

Note further that all of $\triangle ABR$, $\triangle CDS$, and $\triangle EFT$ are also equilateral triangles. Note the side length of $\triangle ABR$ is 10 and the side length of $\triangle CDS$ is 8. Hence the large triangle RST has side length

$$10 + 10 + 8 = 28.$$

This implies that $\triangle EFT$ has side length

$$28 - 8 - 8 = 12.$$

Therefore

$$[ABCDEF] = [RST] - [ABR] - [CDS] - [EFT].$$

Using the formula for area of equilateral triangles, we have

$$[ABCDEF] = \frac{28^2\sqrt{3}}{4} - \frac{10^2\sqrt{3}}{4} - \frac{8^2\sqrt{3}}{4} - \frac{12^2\sqrt{3}}{4} = 119\sqrt{3}.$$

Problem 6.12 Find the area of a regular octagon with side length 12.

Answer

$$288 + 288\sqrt{2}$$

Solution 1

The octagon can be broken down into 4 45-45-90 triangles, 4 rectangles, and 1 square as in the diagram below.

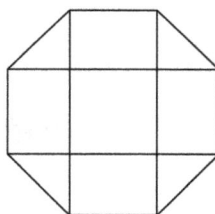

Hence the 45-45-90 triangles have hypotenuse 12 and legs of length

$$12 \times \frac{1}{\sqrt{2}} = \frac{12}{\sqrt{2}} = 6\sqrt{2}.$$

Thus the rectangles have base 12 and height $6\sqrt{2}$ and the square has side length 12. The area of one 45-45-90 triangle is

$$\frac{1}{2} \times 6\sqrt{2} \times 6\sqrt{2} = 36.$$

The area of one rectangle is

$$12 \times 6\sqrt{2} = 72\sqrt{2}.$$

Finally, the area of the square is

$$12^2 = 144.$$

Hence the area of the octagon is

$$144 + 4(72\sqrt{2}) + 4(36) = 288 + 288\sqrt{2}.$$

Solution 2

Adding 4 more 45-45-90 triangles to the above diagram we get a square with side length
$$6\sqrt{2} + 12 + 6\sqrt{2} = 12 + 12\sqrt{2}.$$

(Note this is equivalent to extending every other side of the octagon to form a square.) Hence the full square has area

$$(12 + 12\sqrt{2})^2 = 432 + 288\sqrt{2}.$$

As we know each extra 45-45-90 triangle has area 36 from the previous question, the octagon thus has area

$$432 + 288\sqrt{2} - 4(36) = 288 + 288\sqrt{2}.$$

Problem 6.13 **Mark Y inside regular hexagon $PQRSTU$, so that PQY is equilateral. Is RYU straight?**

Answer

Yes

Solution

Recall a regular hexagon can be broken down into 6 congruent equilateral triangles. Thus $\triangle PQY \cong \triangle PYU \cong \triangle QYR$ and all three are equilateral. Hence
$$\angle RYU = 60° + 60° + 60° = 180°$$

so RYU is straight.

Problem 6.14 Suppose that *EARTH* is a regular pentagon and regular pentagons *ANGLE* and *RAPID* are attached to the outside of the pentagon. Show that N, G, P, I are lie on a single line.

Solution

Note
$$\angle RAP + \angle EAR + \angle NAE = 3 \times 108 = 324°.$$

Hence
$$\angle PAN = 360 - 324 = 36°.$$

As $\triangle PAN$ is isosceles,

$$\angle PNA = \angle NPA = \frac{1}{2}(180 - 36) = 72°.$$

As $72 + 108 = 180°$, we have that both $\angle TPN, \angle GNP$ are straight line angles, and hence all four points are on the same line.

Problem 6.15 A stop sign — a regular *octagon* — can be formed from a square sheet of metal by making four straight cuts that snip off the corners. If we want an octagon with sides of length $\sqrt{2}$, how large does the side of the original square need to be?

Answer

$2 + \sqrt{2}$

Solution

Note the four corners cut off are each 45-45-90 triangles with hypotenuse $\sqrt{2}$. Hence their other sides have length 1. Hence, the side of the square is

$$\sqrt{2} + 1 + 1 = 2 + \sqrt{2}.$$

Problem 6.16 (2009 AMC 8) Construct a square on one side of an equilateral triangle. On one non-adjacent side of the square, construct a regular pentagon, as shown. On one non-adjacent side of the pentagon, construct a hexagon. Continue to construct regular polygons in the same way, until you construct an octagon.

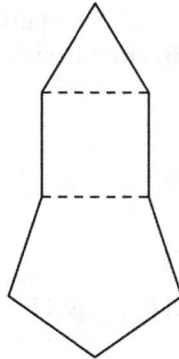

How many sides does the resulting polygon have?

Answer

23

Solution

Note that by combining the polygons described in the problem, the triangle and octagon will share only one side with another polygon, and the other polygons will share two sides. Therefore, the triangle and the octagon contributes to

$$3 + 8 - 2(1) = 9$$

sides of the overall polygon and the remaining polygons contribute to

$$4 + 5 + 6 + 7 - 4(2) = 14$$

sides. Therefore, there are

$$9 + 14 = 23$$

sides to the newly formed polygon.

Problem 6.17 **A triangle has a 60° angle and a 45° angle, and the side opposite the 45° angle has length 12. How long is the side opposite the 60° angle?**

Answer

$6\sqrt{6}$.

Solution

The remaining angle is

$$180° - 60° - 45° = 75°.$$

Draw the altitude from the vertex of $75°$ as in the diagram below

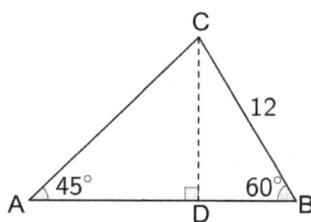

So $\triangle ADC$ is a 45-45-90 triangle and $\triangle BDC$ is a 30-60-90 triangle. Based on the 30-60-90 triangle

$$CD = 12 \times \frac{\sqrt{3}}{2} = 6\sqrt{3},$$

so

$$AC = 6\sqrt{3} \times \sqrt{2} = 6\sqrt{6}$$

based on the 45-45-90 triangle.

Problem 6.18 Let $ABCD$ be a square with area 3 and let EF be a line segment that divides the square into two congruent trapezoids. What is the perimeter of trapezoid $EFBC$?

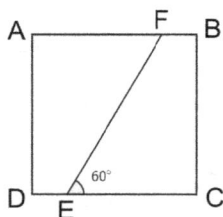

Answer

$2 + 2\sqrt{3}$

Solution

Note that we can form a 30-60-90 triangle as follows:

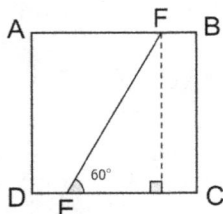

Given that the area of the square is 3, the side length of the square is $\sqrt{3}$. This length also corresponds to the 60° angle in the triangle, which implies that the length from point E to the right angle is 1 and the hypotenuse has length 2.

Since the trapezoids are congruent, we can establish that the shorter base of the trapezoid has length $\frac{\sqrt{3}-1}{2}$ and the longer base of the trapezoid has length $\frac{\sqrt{3}+1}{2}$. Therefore, the perimeter of the trapezoid is

$$\frac{\sqrt{3}+1}{2} + \frac{\sqrt{3}-1}{2} + \sqrt{3} + 2 = 2\sqrt{3} + 2.$$

Problem 6.19 (2015 AMC 8) In the given figure hexagon $ABCDEF$ is equiangular, $ABJI$ and $FEHG$ are squares with areas 18 and 32 respectively, $\triangle JBK$ is equilateral and $FE = BC$.

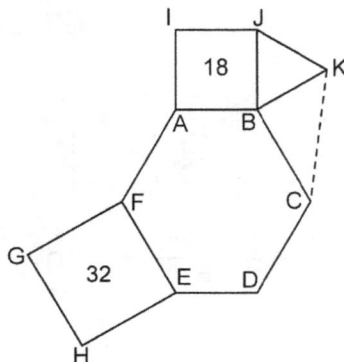

What is the area of $\triangle KBC$?

Answer

12

Solution

Since $\overline{FE}$ is a side of a square with area 32,

$$FE = \sqrt{32} = 4\sqrt{2}.$$

We are given $FE = BC$ so $BC = 4\sqrt{2}$ as well. Similarly, $\overline{JB}$ is a side of a square with area 18, so

$$JB = \sqrt{18} = 3\sqrt{2},$$

and $BK = JB = 3\sqrt{2}$ since $\triangle JBK$ is equilateral. Since an equiangular hexagon has each angle equal to $720° \div 6 = 120°$, we see that

$$\angle JBA + \angle ABC + \angle CBK + \angle KBJ = 360°$$

which implies that

$$90° + 120° + \angle CBK + 60° = 360°,$$

so $\angle CBK = 90°$. Therefore, $\triangle KBC$ is a right triangle with legs $3\sqrt{2}$ and $4\sqrt{2}$ so its area is

$$\frac{3\sqrt{2} \cdot 4\sqrt{2}}{2} = \frac{24}{2} = 12.$$

Problem 6.20 **A stop sign — a regular *octagon* — can be formed from a square sheet of metal by making four straight cuts that snip off the corners. If we have a square with sides of length $\sqrt{2}$, what is the side length of the resulting octagon?**

Answer

$2 - \sqrt{2}$

Solution

Note the four corners cut off are each 45-45-90 triangles. Let x be the side length of the regular octagon. Then, the leg length of the 45-45-90 triangle with hypotenuse length x is

$$x \times \frac{1}{\sqrt{2}} = \frac{x}{\sqrt{2}} = \frac{x\sqrt{2}}{2}.$$

Note the base of the square consists of two of these legs as well as a side of the octagon. Hence,

$$x\sqrt{2} + x = \sqrt{2}.$$

implies that

$$x = \frac{\sqrt{2}}{\sqrt{2}+1} = \frac{\sqrt{2}}{\sqrt{2}+1} \times \frac{\sqrt{2}-1}{\sqrt{2}-1} = 2 - \sqrt{2}$$

Problem 6.21 **Let** ABC **be a triangle given below:**

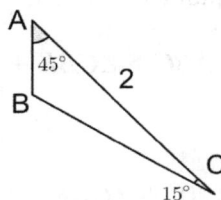

Find the perimeter of the triangle.

Answer

$\frac{1}{3}\sqrt{6} + \sqrt{2} + 2$

Solution

If we extend the line segment AB to point O so that we create a right triangle AOC, we observe that the obtuse triangle is the result of taking a $30 - 60 - 90$ triangle from a $45 - 45 - 90$ triangle. Given that $AC = 2$, by special right triangle side ratios for $45 - 45 - 90$,

$$AO = CO = \sqrt{2}.$$

Note that CO is the side that also corresponds to the 60° in triangle BOC. By using the side ratios for $30 - 60 - 90$ triangles, we have that

$$BO = \sqrt{\frac{2}{3}}$$

and

$$BC = 2\sqrt{\frac{2}{3}}.$$

This implies that

$$AB = \sqrt{2} - \sqrt{\frac{2}{3}}$$

and the perimeter is

$$AB + BC + CA = \sqrt{2} - \sqrt{\frac{2}{3}} + 2\sqrt{\frac{2}{3}} + 2 = \frac{1}{3}\sqrt{6} + \sqrt{2} + 2.$$

Problem 6.22 **From the previous question, determine the area of triangle ABC.**

Answer

$$1 - \frac{\sqrt{3}}{3}$$

Solution

From the previous solution, we have established that

$$AB = \sqrt{2} - \sqrt{\frac{2}{3}}$$

and $CO = \sqrt{2}$. Therefore, the area of ABC is

$$\frac{1}{2} \times \sqrt{2} \times \left(\sqrt{2} - \sqrt{\frac{2}{3}} \right) = 1 - \frac{\sqrt{3}}{3}.$$

Problem 6.23 **Let ABC be an equilateral triangle that contains a 45-45-90 triangle DEF with hypotenuse parallel to a side of the equilateral triangle given below:**

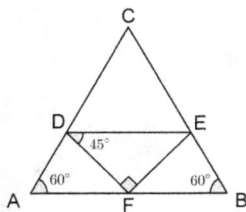

If $AB = 1$, find DF.

Answer

$$\frac{1}{4}(3\sqrt{2} - \sqrt{6})$$

Solution

Note that since the hypotenuse $DE \parallel AB$, this implies that CDE is also equilateral. If we let x represent the length of DF, then CDE has side length $x\sqrt{2}$.

Furthermore, note that ADF can be redrawn as a $30 - 60 - 90$ triangle with a $45 - 45 - 90$ triangle. By special right triangles, the height of ADF is $\frac{x}{\sqrt{2}}$.

The height also corresponds to the $60°$ of the created $30 - 60 - 90$ triangle. This implies that

$$AD = \frac{2x}{\sqrt{6}}.$$

Given that $AC = 1$, we have

$$AC = AD + DC = x\sqrt{2} + \frac{2x}{\sqrt{6}} = 1.$$

Solving for x yields $DF = \frac{1}{4}(3\sqrt{2} - \sqrt{6})$.

Problem 6.24 **Using the above figure, find the area of triangle DEF.**

Answer

$$\frac{3}{8}(2 - \sqrt{3})$$

Solution

Taking the previous result, the area of DEF is

$$\frac{1}{2}(\frac{1}{4}(3\sqrt{2}-\sqrt{6}))^2 = \frac{3}{8}(2-\sqrt{3}).$$

Problem 6.25 **Using the above figure, find the side ratio of triangle DAF.**

Answer

$2 : \sqrt{6} : \sqrt{3}+1$ w.r.t angle ratio $45 - 60 - 75$

Solution

Without loss of generality, let $DF = \sqrt{2}$. Recall that triangle ADF can be decomposed into $30 - 60 - 90$ and $45 - 45 - 90$ triangles. The side length of the $45 - 45 - 90$ triangle formed by connecting D to line segment AF is 1. Again, the length 1 also corresponds to the $60°$ in the $30 - 60 - 90$ triangle.

Therefore, the side corresponding to the $30°$ has length equal to $\frac{1}{\sqrt{3}}$ and hypotenuse AD has length $\frac{2}{\sqrt{3}}$.

Thus, $AF = \frac{1}{\sqrt{3}} + 1$ and $AD : DF : AF$ has side ratios $2 : \sqrt{6} : \sqrt{3}+1$.

7. Discovering Areas

Problem 7.1 Let $ABCD$ be a square and let $EFGH$ be a rectangle such that $[ABCD] = [EFGH]$. **Suppose that the side length of the rectangle is twice the side length of the square. Find the ratio of the perimeter of the square to the perimeter of the rectangle.**

Answer

$4:5$

Solution

Let x be the length of square $ABCD$. Then, the side length of rectangle $EFGH$ is $2x$ and the width of rectangle $EFGH$ is $\frac{x}{2}$. Therefore, the perimeter of the square is $4x$ and the perimeter of the rectangle is

$$2x + 2x + \frac{1}{2}x + \frac{1}{2}x = 5x.$$

The ratio of the perimeter of the square to the perimeter of the rectangle is $4:5$.

Problem 7.2 **What is the area of a regular hexagon with side length** 3**?**

Answer

$\dfrac{27}{2}\sqrt{3}$

Solution

A regular hexagon can be decomposed into 6 equilateral triangles with side length 3. The area of one equilateral triangle (using $30 - 60 - 90$) is

$$\frac{1}{2} \times 3 \times \frac{3}{2}\sqrt{3} = \frac{9}{4}\sqrt{3}.$$

Therefore, the area of a regular hexagon with side length 3 is $\frac{27}{2}\sqrt{3}$.

Problem 7.3 **What is the area of a regular octagon with side length** 3**?**

Answer

$18(1 + \sqrt{2})$

Solution

A regular octagon can be reinterpreted as a square minus four $45 - 45 - 90$ triangles. Specifically, the area of a regular octagon with side length 3 is

$$\left(3 + 2\left(\frac{3}{\sqrt{2}}\right)\right)^2 - 2\left(\frac{3}{\sqrt{2}}\right)^2 = 18(1 + \sqrt{2}).$$

Problem 7.4 **If the area of parallelogram** $ABCD$ **is** 10 **and the area of triangle** CDE **is** 2**, what is the area of trapezoid** $ABCE$**?**

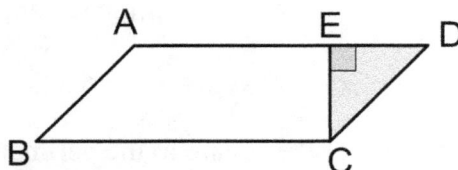

Answer

8

Solution

Note that
$$[ABCE] = [ABCD] - [CDE] = 10 - 2 = 8.$$

Problem 7.5 If the area of the rectangle is 40, the area of the smaller white triangle is 5, and the area of the big white triangle is 20, what is the area of the shaded region?

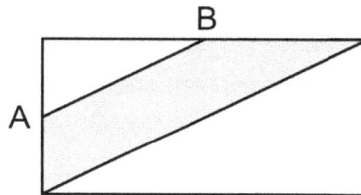

Answer

15

Solution

The area of the shaded region is
$$40 - 5 - 20 = 15.$$

Problem 7.6 Suppose you make a parallelogram with base 5cm and height 4cm.

(a) Find the area of the parallelogram.

Answer

20

Solution

The base is 5 and the height 4, so the area is

$$5 \times 4 = 20 \text{cm}.$$

(b) **Show it is possible to have different perimeters for a parallelogram as above. Is there a largest perimeter that is possible? What about a smallest?**

Answer

Smallest is 18

Solution

If the parallelogram is a rectangle the perimeter is

$$2 \times 5 + 2 \times 4 = 18 \text{cm}.$$

If it is a parallelogram, the perimeter is larger, and the more "stretched out" the parallelogram is, the larger the perimeter, so there is no largest perimeter.

Problem 7.7 **The parallelogram $ABCD$ has area 300cm^2 and E is the midpoint of $\overline{AD}$ Find the area of the shaded region.**

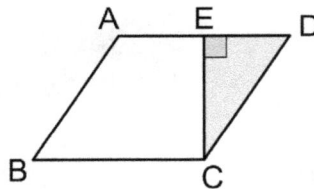

Answer

75

Solution

As the two triangles share the same height and E is a midpoint,

$$[ADC] = 2[EDC].$$

Then

$$[ABCD] = 2[ADC]$$

as the triangle and parallelogram share the same height and base. Hence

$$[EDC] = \frac{1}{4}[ABCD] = 75.$$

Problem 7.8 **A garden of rectangular shape is shown in the diagram. The shaded regions are grass, and the unshaded regions are empty spaces in the shape of six congruent rhombi. Find the ratio between the areas of the grass and empty regions.**

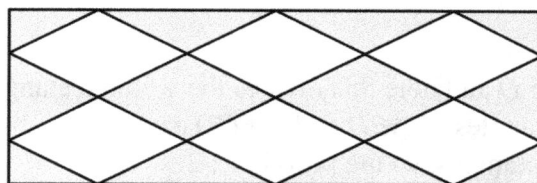

Answer

1 : 1

Solution

Note that the rectangle consists of 6 unshaded rhombi, 2 shaded rhombi, 6 shaded half-rhombi, and 4 shaded quarter-rhombi. Further, each of these rhombi is congruent. Hence, if each rhombi has area 1, The ratio of un-shaded to shaded is

$$6 : \left(2 + 4 \times \frac{1}{2} + 4 \times \frac{1}{4}\right) = 4 : 4 = 1 : 1.$$

Problem 7.9 **In the diagram, points A, B, C, D are the midpoints of their respective sides. Compute the ratio of the area of the shaded region and the whole rectangle.**

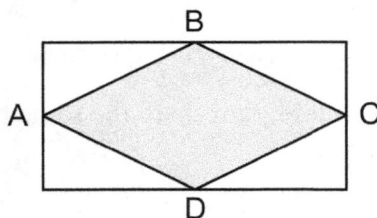

1 : 2

Solution

Connect B and D and note this divides the whole rectangle into two smaller congruent rectangles. $\triangle ABD$ and $\triangle BCD$ each share a base and height with the smaller rectangles, so the ratio is 1 : 2.

Problem 7.10 **In the parallelogram** $ABDC$, **points** E **and** F **are the midpoints of sides** $\overline{AD}$ **and** $\overline{DC}$ **respectively. Which triangles have the same area as** $\triangle BFC$**?**

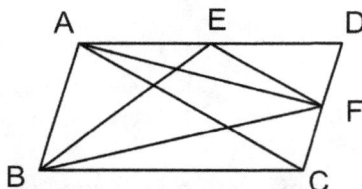

Solution

The area of $\triangle BFC$ is $\dfrac{1}{4}$ of the parallelogram. There are 3: $\triangle AFC$, $\triangle ADF$, $\triangle ABE$.

Problem 7.11 **In the rectangle** $ABCD$, **the area of** $\triangle AOB$ **is** 6 cm^2. **Let** O **be a point inside rectangle such that the area of** $\triangle DOC$ **is** 1/3 **of the area of the rectangle. Find the area of rectangle** $ABCD$.

Answer

36

Solution

The areas of $\triangle AOB$ and $\triangle DOC$ add up to half of the area of the rectangle $ABCD$. Therefore

$$[AOB] = \frac{1}{6}[ABCD].$$

Thus the rectangle area is

$$6 \times 6 = 36\text{cm}^2.$$

Problem 7.12 (2015 AMC 8) Point O is the center of the regular octagon $ABCDEFGH$, and X is the midpoint of the side $\overline{AB}$.

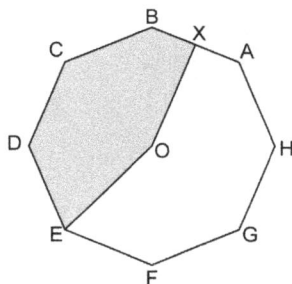

What fraction of the area of the octagon is shaded?

Answer

$\dfrac{7}{16}$

Solution

Since the octagon is regular we know that

$$\triangle AOB \cong \triangle BOC \cong \triangle COD \cong \cdots \cong \triangle HOA,$$

and each has area $\frac{1}{8}$th of the octagon. Further, as X is the midpoint of AB,

$$[XOB] = \frac{1}{2}[AOB].$$

Thus the shaded region has area $3.5[AOB]$ so the shaded region is

$$\frac{3.5}{8} = \frac{7}{16}$$

of the entire octagon.

Problem 7.13 **Let $ABCD$ be a parallelogram, as in the diagram. Compare the shaded regions $\triangle ABF$ and $\triangle DEF$, which one has the larger area?**

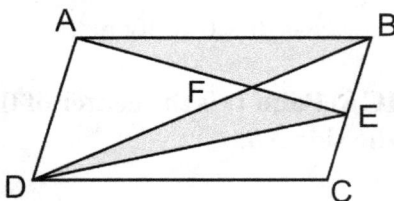

Answer

They have the same area

Solution

Note $[ABD] = [DEF]$ as they have the same height and base. Hence, after removing the shared $\triangle AFD$ we see the two shaded triangles have the same area.

Problem 7.14 **In the diagram, points A and B are the midpoints of their respective sides. Compute the ratio of the area of the shaded region and the whole rectangle.**

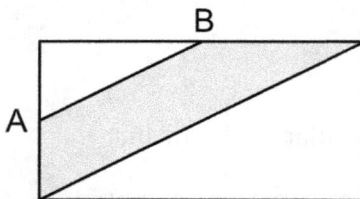

Answer

3 : 8

Solution

Connect A and B and the midpoint of the diagonal, and it is clear that this divides the upper half of the rectangle into 4 congruent triangles. Hence $\frac{3}{4}$ of the upper half is shaded so

$$\frac{1}{2} \times \frac{3}{4} = \frac{3}{8}$$

of the entire rectangle is shaded. As a ratio this is 3 : 8.

Problem 7.15 (2005 AMC 8) The area of polygon $ABCDEF$ is 52 with $AB = 8$, $BC = 9$ and $FA = 5$. What is $DE + EF$?

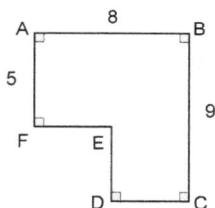

Answer

9

Solution

Firstly, note that
$$AF + DE = BC,$$

so $DE = 4$. If we extend the length of segments DC and AF, the segments will intersect at point O which creates the rectangular region $ABCO$.

Therefore, the area of the polygon can be determined by taking the area of $FEDO$ and subtracting it from the area of $ABCO$. Therefore,

$$52 = 8 \times 9 - 4 \times EF$$

implies that $EF = 5$. The answer is

$$DE + EF = 4 + 5 = 9.$$

Problem 7.16 **A garden of rectangular shape is shown in the diagram. The shaded regions are grass, and the unshaded regions are empty spaces in the shape of four congruent hexagons. Find the ratio between the areas of the grass and empty regions.**

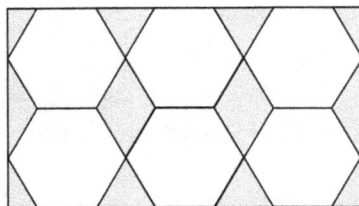

Answer

     $1 : 3$

Solution

Recall that a hexagon can be broken down into 6 equilateral triangles. From here we can see that the unshaded region has the area of

$$6 \times 6 = 36$$

of these equilateral triangles. The shaded region has the area of

$$2 \times 2 + 4 \times 1 + 8 \times \frac{1}{2} = 12$$

equilateral triangles. Hence the ratio of $12 : 36 = 1 : 3$.

Problem 7.17 **In the diagram, $\triangle ABC, \triangle DEF$ are two congruent isosceles right triangles. Given that $AB = 9, EC = 3$, find the area of the shaded region.**

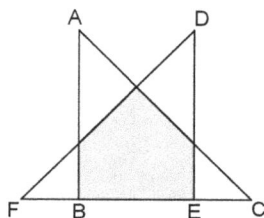

Answer

27

Solution

Label the resulting pentagon $BEGHI$. We have

$$AB = BC = 9,$$

so

$$BE = 9 - 3 = 6.$$

Therefore, $FB = 3$ so $CF = 12$. Now note $\triangle CFH$ is a 45-45-90 triangle with hypotenuse 12. Hence its sides are length $\frac{12}{\sqrt{2}}$, so it has area

$$[CFH] = \frac{1}{2} \times \frac{12}{\sqrt{2}} \times \frac{12}{\sqrt{2}} = 36.$$

Similarly,

$$[CEG] = [FBI] = \frac{1}{2} \times 3 \times 3 = \frac{9}{2}.$$

Hence, the shaded pentagon has area

$$36 - 9 = 27.$$

Problem 7.18 (**2000 AMC 8**) **Triangles** ABC, ADE, **and** EFG **are all equilateral. Points** D **and** G **are midpoints of** $\overline{AC}$ **and** $\overline{AE}$, **respectively. If** $AB = 4$, **what is the perimeter of figure** $ABCDEFG$?

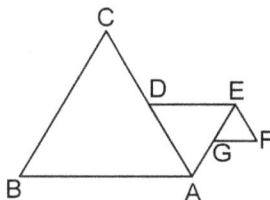

Answer

15

Solution

Since $AB = 4$, the large triangle ABC has sides of length 4.

It follows that the medium triangle has sides of length 2 and the small triangle has sides of length 1.

In the figure, there are 3 segment sizes, and all segments depicted in the figure are one of these lengths. Therefore, the perimeter is

$$AB + BC + CD + DE + EF + FG + GA = 4 + 4 + 2 + 2 + 1 + 1 + 1 = 15.$$

Problem 7.19 In the given trapezoid $ABCD$, there are 8 triangles. Among them, the pair $\triangle ABC$ and $\triangle DBC$ have the same area. How many other pairs have the same areas? List the pairs with the same areas.

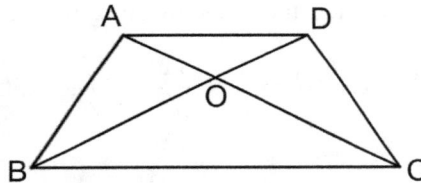

Answer

2 more pairs

Solution

There are 2 more. $\triangle ABD, \triangle ACD$, and $\triangle ABO, \triangle DCO$.

Problem 7.20 (2011 AMC 8) Quadrilateral $ABCD$ is a trapezoid, $AD = 15$, $AB = 50$, $BC = 20$, **and the altitude is** 12.

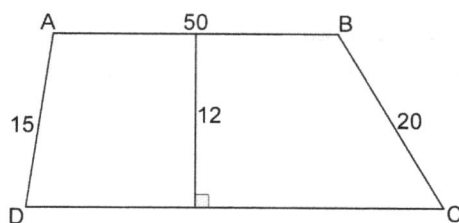

What is the area of the trapezoid?

Answer

750

Solution

If you draw altitudes from AX and BX to CD, the trapezoid will be divided into two right triangles and a rectangle.

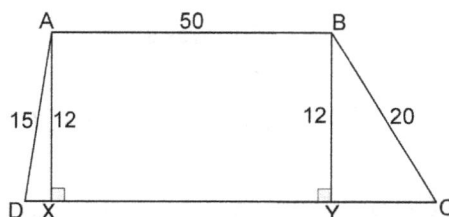

Using Pythagorean theorem, the values of DX and CY can be computed as follows:

$$DX = \sqrt{15^2 - 12^2} = \sqrt{81} = 9$$

$$CY = \sqrt{20^2 - 12^2} = \sqrt{256} = 16$$

Since $ABYX$ is a rectangle, $XY = AB = 50$. Therefore,

$$CD = DX + XY + CY = 9 + 50 + 16 = 75$$

The area of the trapezoid is

$$12 \times \frac{(50 + 75)}{2} = 6 \times 125 = 750.$$

Problem 7.21 Let $ABCD$ **be a parallelogram with** $[ABCD]$**. Let** P **be a point in the interior of** $ABCD$**. Show that** $[ABP] + [CDP] = [ABCD]/2$**.**

Solution 1

Construct line segment $\overline{EF}$ with E, F on $\overline{AD}, \overline{BC}$ such that $\overline{EF}$ goes through P and $\overline{EF} \parallel \overline{AB}, \overline{CD}$.

We then have:
$$[CDP] = [CDF]$$
and
$$[ABP] = [ABF] = [DBF]$$
(draw these out!). Hence,

$$[ABP] + [CDP] = [DBF] + [CDF] = [BCD] = \frac{1}{2}[ABCD].$$

Solution 2

Note the height from $\overline{AB}$ to $\overline{CD}$ through P splits into altitudes for the two given triangles. As these altitudes sum to the total height of the parallelogram,

$$[ABP] + [CDP] = \frac{1}{2}[ABCD].$$

Problem 7.22 Let $ABCD$ **be a parallelogram, as in the diagram. Suppose** E **is the midpoint of** $\overline{BC}$**. Find the area of the shaded regions** $\triangle ABF$ **and** $\triangle DEF$ **in terms of the entire area of the parallelogram.**

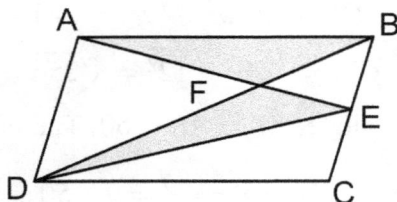

Answer

$\dfrac{1}{3}[ABCD]$

Solution

First note (since E is a midpoint) that

$$[DEC] = [BED] = \frac{1}{4}[ABCD].$$

Since $ABCD$ is a parallelogram and $BE = \frac{1}{2}AD$, $\triangle AFD \sim \triangle EFB$, with ratio of side lengths $2 : 1$. Using this information, we have

$$[DFE] = 2[BEF].$$

Therefore,

$$[DFE] = [ABF] = \frac{1}{6}[ABCD].$$

Hence the sum of the shaded regions is $\frac{1}{3}[ABCD]$.

Problem 7.23 **If regular hexagon $ABCDEF$ with side length 2 can be reinterpreted as 6 equilateral triangles, what is the area of $ABCDEF$?**

Answer

$6\sqrt{3}$

Solution

Note that $ABCDEF$ can be decomposed into 6 equilateral triangles each with side length 2.

We can apply the rules of special right triangles to deduce that the height of one equilateral triangle is $\sqrt{3}$. The area of one equilateral triangle is

$$\frac{1}{2} \times 2 \times \sqrt{3} = \sqrt{3}.$$

Since there are six equilateral triangles in $ABCDEF$, $[ABCDEF] = 6\sqrt{3}$.

Problem 7.24 **(2006 AMC 8) The letter T is formed by placing two 2×4 inch rectangles next to each other, as shown. What is the perimeter of the T, in inches?**

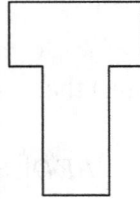

Answer

20

Solution

Individually, the perimeter of the each rectangle is

$$2 \times (2+4) = 12.$$

Since their connection removes a total of 4 units from the perimeter of the figure, we observe that the final perimeter is

$$2 \times 12 - 4 = 20.$$

Problem 7.25 **Find the perimeter of the quadrilateral formed by the coordinates** $A = (0,0)$, $B = (4,3)$, $C = (6,3)$, **and** $D = (10,0)$?

Answer

32

Solution

Observe that by plotting the points, we create a trapezoid consisting of two $3-4-5$ triangles and a 3×2 rectangle. Note that

$$AB = \sqrt{3^2 + 4^2} = 5,$$

and

$$CD = \sqrt{3^2 + 4^2} = 5.$$

Also, note that $BC = 2$ and $AD = 10$. Therefore, the perimeter of the trapezoid is

$$10 + 2 + 5 + 5 = 32.$$

8. Conquering Areas

Problem 8.1 Two people are building (tiny) houses. Both people will have triangular roofs that are 5m long (for each half), with a square bottom of the house. The first plans to have a house 8m wide, and the second plans to have a house 6m wide, as shown below:

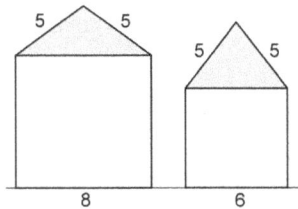

How tall is each house?

Answer

11, 10

Solution

Note that the shaded roof space can be divided into two right triangles in each case. Using the Pythagorean theorem, we see that both triangles are $3, 4, 5$ triangles. Therefore, the roofs have heights 3 and 4, so the houses are

$$8 + 3 = 11$$

or

$$6 + 4 = 10$$

meters tall.

Problem 8.2 **Using the above diagram, find the amount of attic space in each house. (That is the shaded region.)**

Answer

12

Solution

The first house's attic is a triangle with base 8 and height 3, so has area

$$\frac{1}{2} \times 8 \times 3 = 12.$$

The second house's attic is a triangle with base 6 and height 4, so has area

$$\frac{1}{2} \times 6 \times 4 = 12.$$

Problem 8.3 **Find the total amount of space in each house (including the attic).**

Answer

76, 48.

Solution

Both houses have attics of size 12. Adding this to the square bottoms give total areas of

$$8^2 + 12 = 76$$

and

$$6^2 + 12 = 48.$$

Problem 8.4 Suppose we have a right triangle such that $AC = 3$, $BC = 4$, and $\angle C = 90°$. What is the area of this triangle?

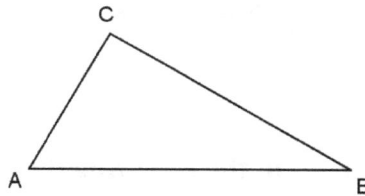

Answer

6

Solution

BC can be reinterpreted as the base of the right triangle and AC can be reinterpreted as the height of the right triangle. Therefore, the area is

$$\frac{1}{2} \times 3 \times 4 = 6.$$

Problem 8.5 Now suppose we have the same right triangle such that $AC = 3$, $BC = 4$, and $\angle C = 90°$. What is the length of altitude CD?

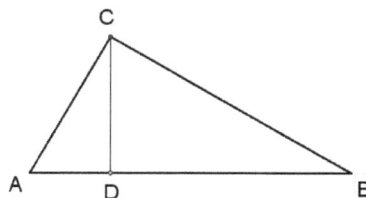

Answer

$$\frac{12}{5}$$

Solution

By the previous problem, the area of the triangle is 6. Since $3 - 4 - 5$ is a Pythagorean triple, we have $AB = 5$. Therefore,

$$6 = \frac{1}{2} \times 5 \times CD$$

or $CD = \frac{12}{5}$.

Problem 8.6 All the rectangles in the following diagrams are squares, except for (d), where the triangles are isosceles right triangles. The lengths of the segments are marked. Find the area of the shaded regions in each diagram.

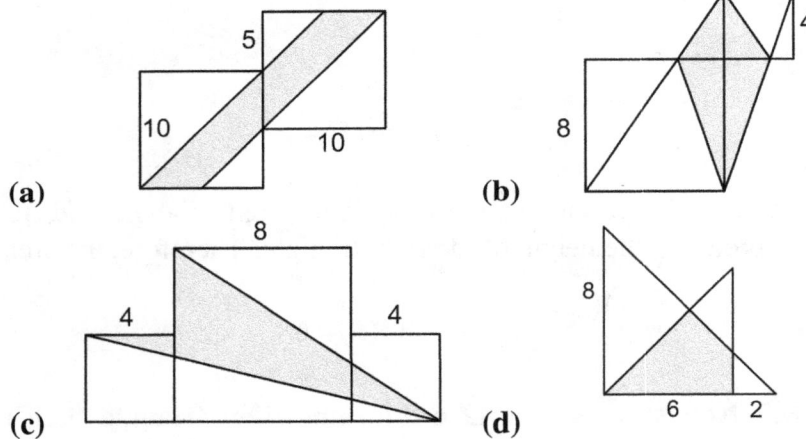

(a)

(b)

(c)

(d)

Answer

(a) 75, (b) 32, (c) 32, (d) 14

Solution

For (a), the shaded region is a parallelogram with base $10 - 5 = 5$ and height $10 + 5 = 15$, hence has area

$$15 \times 5 = 75.$$

For (b), consider the following diagram:

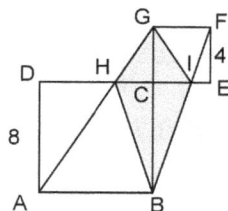

Hence the shaded region is

$$[ABFG] - [ABH] - [FGI].$$

$ABFG$ is a trapezoid, hence has area

$$\frac{1}{2}(8 + 4) \times 12\frac{1}{2} = 72.$$

We have

$$[ABH] = \frac{1}{2} \times 8 \times 8 = 32, [FGI] = \frac{1}{2} \times 4 \times 4 = 8$$

so $[GHBI] = 72 - 32 - 8 = 32$.

For (c), first extend lines to make a big rectangle with height 8 and length 16. The shaded region is then the area of the big rectangle, minus a 4 by 4 square and two triangles (with respective bases and heights $16, 4$ and $12, 8$. The area of the shaded region is then

$$16 \times 8 - 4^2 - \frac{1}{2} \times 16 \times 4 - \frac{1}{2} \times 12 \times 8 = 128 - 16 - 32 - 48 = 32.$$

For (d), note all the triangles are isosceles right triangles. The area of the triangle with base and height 6 is hence

$$\frac{1}{2} \times 6 \times 6 = 18.$$

This is exactly the shaded region plus a little extra. The little extra is an isosceles right triangle with hypotenuse 4. Therefore the base and height of this triangle are

$$4 \times \frac{1}{\sqrt{2}} = \frac{4}{\sqrt{2}} = 2\sqrt{2},$$

so has area

$$\frac{1}{2} \times 2\sqrt{2} \times 2\sqrt{2} = 4.$$

Hence the shaded region has area $18 - 4 = 14$.

Problem 8.7 **In parallelogram** $ABCD$, M **and** N **are midpoints of sides** $\overline{AB}$ **and** $\overline{BC}$ **respectively. Given that** $[DMN] = 9$, **find the area of** $ABCD$.

Answer

24

Solution

It is easy to find that

$$[AMD] = [CDN] = \frac{1}{4}[ABCD],$$

and

$$[BMN] = \frac{1}{8}[ABCD].$$

Therefore

$$[DMN] = \frac{3}{8}[ABCD],$$

so $[ABCD] = 24$.

Problem 8.8 **In the diagram below, there are** 36 **rectangular grid points, evenly spaced, and the distance between each pair of adjacent points is** 1. **Find the area of quadrilateral** $ABCD$.

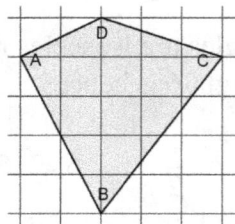

Answer

$$\frac{25}{2}$$

Solution

Note that $\overline{AC}, \overline{BD}$ divides the entire square into four rectangles. For each of those rectangles, note that half of the rectangle is shaded, hence the area of $ABCD$ is half of the area of the entire grid, so

$$[ABCD] = \frac{5^2}{2} = \frac{25}{2}.$$

Problem 8.9 **In the diagram below, there are** 21 **grid points arranged in equilateral triangles, equally spaced. The** *area* **of each small equilateral triangle formed by** 3 **adjacent grid points is** 1. **Find the area of quadrilateral** *ABCD*.

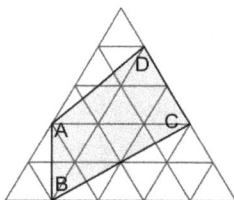

Answer

12

Solution

We subtract the areas of the unshaded triangles from the total area (which is 25). The triangle above $\overline{AD}$ is half of a parallelogram of area 6, so has area 3.

The triangle to the right of $\overline{AB}$ is half a triangle of area 4, so has area 2.

Lastly, the triangle below $\overline{BC}$ is half a parallelogram of area 16, hence has area 8. Hence

$$[ABCD] = 25 - 3 - 2 - 8 = 12.$$

Problem 8.10 (**2015 AMC 8**) **A triangle with vertices as** $A = (1,3)$, $B = (5,1)$, **and** $C = (4,4)$ **is plotted on a** 6×5 **grid.**

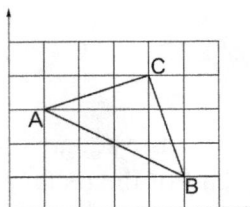

What fraction of the grid is covered by the triangle?

Answer

$\dfrac{1}{6}$

Solution

First note the 6×5 grid is a rectangle so it has area

$$6 \times 5 = 30.$$

For $\triangle ABC$, first remove the outer grid, leaving the grid containing $\triangle ABC$ in the diagram below.

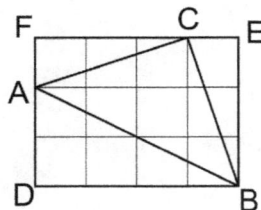

This remaining grid is a 4×3 rectangle $BEFG$. Note that

$$[ABC] = [BEFD] - [BEC] - [CFA] - [ADB].$$

Hence

$$[ABC] = 4 \times 3 - \frac{1}{2} \times 3 \times 1 - \frac{1}{2} \times 3 \times 1 - \frac{1}{2} \times 4 \times 2 = 12 - 1.5 - 1.5 - 4 = 5.$$

Thus the fraction of the entire grid covered is

$$\frac{5}{30} = \frac{1}{6}.$$

Problem 8.11 **Suppose** $\triangle ABC$ **with** D **on** $\overline{AB}$, G **on** $\overline{AC}$, **and** E, F **on** $\overline{BC}$. **Suppose further** $AD = \frac{1}{2}AB, BE = EF = FC, CG = \frac{1}{2}GA$. **If the area of** $[ABC] = 36$, **find the area of the quadrilateral** $DEFG$.

Answer

15

Solution

Clearly

$$[DEFG] = [ABC] - [DBE] - [FCG] - [ADG].$$

We have

$$[DBE] = \frac{1}{3}[DBC], [DBC] = \frac{1}{2}[ABC]$$

so

$$[DBE] = \frac{1}{6}[ABC] = 6.$$

Similarly,

$$[FCG] = \frac{1}{6}[ABC] = 6$$

as well. Lastly,

$$[ADG] = \frac{1}{2}[ADC], [ADC] = \frac{1}{2}[ABC]$$

so

$$[ADG] = \frac{1}{4}[ABC] = 9.$$

Thus,

$$[DEFG] = 36 - 6 - 6 - 9 = 15.$$

Problem 8.12 **In the square shown in the diagram, the side length is 6, and the sum of the areas of the two shaded regions is 12. Find the area of quadrilateral** *ABCD*.

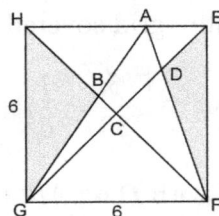

Answer

3

Solution

Note that

$$[AGF] = \frac{1}{2} \times 6 \times 6 = 18,$$

so

$$[AGF] + [HBG] + [DEF] = 18 + 12 = 30.$$

Also the area of the pentagon $GHCEF$ is $\frac{3}{4}$ of $[EFGH]$, which means

$$[GHCEF] = \frac{3}{4} \times 6 \times 6 = 27.$$

Therefore

$$[ABCD] = [AGF] + [HBG] + [DEF] - [GHCEF] = 30 - 27 = 3.$$

Problem 8.13 A square is formed by putting 4 congruent isosceles right triangles at the corners. The shaded square is the region not covered by the triangles. Find the area of the shaded square, if

(a) **As shown in the diagram below, the triangles just touch.**

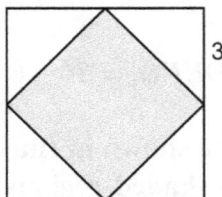

Answer

18.

Solution

Each triangle is a 45-45-90 triangle, so the hypotenuse of the triangle is $3\sqrt{2}$.

As this is the side length of the triangle, the area is 18.

(b) **As shown in the diagram below, the triangles overlap a little. (Note: In this case, 3 is *not* the side length of the isosceles triangle.)**

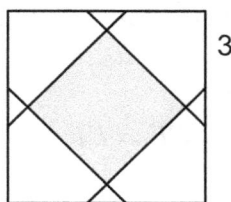

Answer

18.

Solution

Note the opposite hypotenuses of the triangles are parallel. Hence, if we create an isosceles triangle with side length 3, we see its hypotenuse has the same length as the square.

Hence, the square has side length $3\sqrt{2}$ and area 18.

Problem 8.14 In parallelogram $ABCD$ as shown, $BC = 12$. Triangle BCE is a right triangle where $\overline{BE}$ is the hypotenuse, and $EC = 9$. If $AE = CD$, what is the area of the full figure (the pentagon $BCDFE$)?

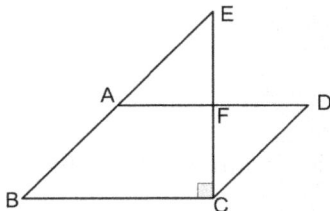

Answer

$$\frac{135}{2}$$

Solution

Recall that $(3, 4, 5)$ is a Pythagorean triple, so $(9, 12, 15)$ is also a Pythagorean triple. Hence $BE = 15$. Further, we know

$$AE = CD = AB,$$

so A is the midpoint of $\overline{BE}$. Since $\triangle AEF \sim \triangle BEC$ using AAA, we know that F is also the midpoint of $\overline{EC}$ so

$$CF = EF = 9 \div 2 = \frac{9}{2}.$$

We also have that F is the midpoint of $\overline{AD}$ so

$$AF = DF = BC \div 2 = 12 \div 2 = 6.$$

Hence the area of the full figure is the parallelogram $ABCD$ plus the area of triangle AFE which is

$$BC \times CF + \frac{1}{2} \times AF \times EF = 12 \times \frac{9}{2} + \frac{1}{2} \times 6 \times \frac{9}{2} = 54 + \frac{27}{2} = \frac{135}{2}.$$

Problem 8.15 Using the above figure (not drawn to scale), if $BE = 10$, $AD = 8$ and the area of right triangle BCE is equal to the area of parallelogram $ABCD$, what is the area of trapezoid $ABCF$?

Answer

18

Solution

Since $ABCD$ is a parallelogram, we have that

$$8 = AD = BC.$$

Given right triangle BCE, by the Pythagorean Theorem, we have $CE = 6$. The area of the triangle is

$$\frac{1}{2} \times 6 \times 8 = 24,$$

and since the area of the parallelogram is equal to the area of the triangle, we get

$$CF = FE = 3.$$

By similar triangles, $AF = 4$. The area of trapezoid $ABCF$ is

$$\frac{1}{2} \times 3 \times (4+8) = 18.$$

Problem 8.16 (2014 AMC 8) Six rectangles each with a common base width of 2 have lengths of $1, 4, 9, 16, 25,$ and 36. What is the sum of the areas of the six rectangles?

Answer

182

Solution

The sum of the areas of the rectangles is equal to

$$2 \times 1 + 2 \times 4 + 2 \times 9 + 2 \times 16 + 2 \times 25 + 2 \times 36 =$$
$$2 \times (1 + 4 + 9 + 16 + 25 + 36) = 2 \times 91 = 182.$$

Problem 8.17 As shown in the diagram, square $ABCD$ has side length 5. Let E and F be the midpoints of $\overline{AB}$ and $\overline{BC}$ respectively. Find the area of quadrilateral $BFGE$.

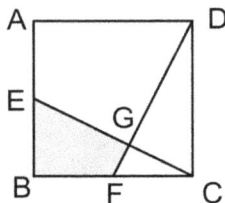

Answer

> 5

Solution

> Connect the other two midpoints to vertices as shown:

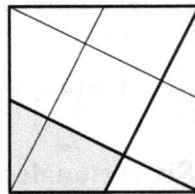

> Note the piece can be rearranged to form 5 congruent squares.
>
> The area of the big square is the sum of the areas of the 5 small squares, so each of the 5 small squares has area
>
> $$5^2 \div 5 = 5.$$
>
> The area of the shaded region is the same as one small square, so the answer is 5.

Problem 8.18 **In the diagram, $\triangle ABC$ and $\triangle DEF$ are two congruent isosceles right triangles. Given that $AB = 9, EC = 3$, find the area of the shaded region.**

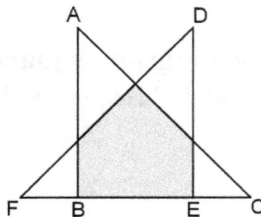

Answer

> 27

Solution

Label the resulting pentagon $BEGHI$. We have

$$AB = BC = 9,$$

so

$$BE = 9 - 3 = 6.$$

Therefore, $FB = 3$ so $CF = 12$. Now note $\triangle CFH$ is a 45-45-90 triangle with hypotenuse 12. Hence its sides are length $\frac{12}{\sqrt{2}}$, so it has area

$$[CFH] = \frac{1}{2} \times \frac{12}{\sqrt{2}} \times \frac{12}{\sqrt{2}} = 36.$$

Similarly,

$$[CEG] = [FBI] = \frac{1}{2} \times 3 \times 3 = \frac{9}{2}.$$

Hence, the shaded pentagon has area $36 - 9 = 27$.

Problem 8.19 In the figure containing right triangle BCE and parallelogram $ABCD$ shown below, $EF = 3$, $EG = 5$ and $BG = 20$. **Find the area of the shaded region.**

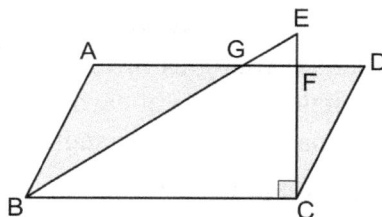

Answer

96

Solution

Given right triangle BCE and parallelogram $ABCD$, we have that EFG is also a right triangle. This implies that

$$FG = \sqrt{5^2 - 3^2} = \sqrt{25 - 9} = \sqrt{16} = 4$$

by the Pythagorean theorem. Note $BE = 20 + 5 = 25$ so $BC = \frac{4}{5} \times 25 = 20$ by similar triangles. Therefore,

$$CE = \sqrt{25^2 - 20^2} = \sqrt{625 - 400} = \sqrt{225} = 15$$

by the Pythagorean theorem, so $CF = 15 - 3 = 12$. The area of the shaded region is the parallelogram $ABCD$ minus the trapezoid $BCFG$ so the shaded region has area

$$(20 \times 12) - \frac{1}{2} \times 12 \times (20 + 4) = 96.$$

Problem 8.20 (2012 AMC 8) An equilateral triangle and a regular hexagon have equal perimeters. If the area of the triangle is 4, what is the area of the hexagon?

Answer

6

Solution

Let s be the side length of an equilateral triangle. Then, for the hexagon to maintain the same perimeter, the regular hexagon must have side length $\frac{s}{2}$.

Note that a hexagon can be decomposed into six equilateral triangles. Since each of the six equilateral triangles have side length that is half of the original equilateral triangle, the area of each of the six equilateral triangles is $\frac{1}{4}$ the area of the original equilateral triangle.

Therefore, each of the six equilateral triangles have area 1, which implies that the area of the hexagon is 6.

Problem 8.21 (2006 AMC 8) Points A, B, C and D are midpoints of the sides of the larger square. If the larger square has area 60, what is the area of the smaller square?

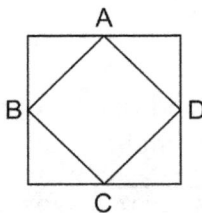

Answer

30

Solution 1

The large square has area 60, so side length

$$\sqrt{60} = 2\sqrt{15}.$$

Note the four triangles in the picture are congruent 45-45-90 triangles, with legs of length $\sqrt{15}$. Hence

$$AB = BC = CD = DA = \sqrt{15} \times \sqrt{2} = \sqrt{30}.$$

Thus the area of $ABCD$ is $\sqrt{30}^2 = 30$.

Solution 2

Connect A to C and B to D. Note this divides the larger square into 8 congruent 45-45-90 triangles. Since half make up the smaller square, $ABCD$ has area $60 \div 2 = 30$.

Problem 8.22 In the diagrams below, the figures are generated by taking the middle third of every side of the previous figure and forming equilateral triangles from the thirds.

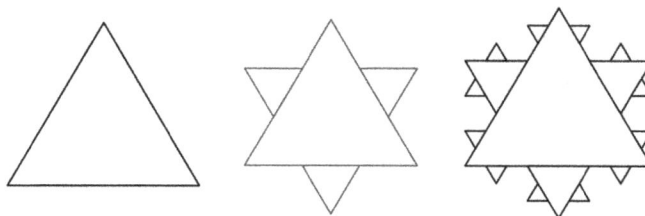

(a) **Let's start with the equilateral triangle! If the equilateral triangle has side length 9, what is the area of the equilateral triangle?**

Answer

$$\frac{81}{4}\sqrt{3}$$

Solution

As we have done before we can use $30 - 60 - 90$ triangles to determine the height of the equilateral triangle with side length 9. By the side ratios of $30 - 60 - 90$ triangles, the height has length

$$9 \times \frac{\sqrt{3}}{2} = \frac{9\sqrt{3}}{2}$$

so the area of the equilateral triangle is

$$\frac{1}{2} \times 9 \times \frac{9\sqrt{3}}{2} = \frac{81}{4}\sqrt{3}.$$

(b) **What is the area of the second figure?**

Answer

$27\sqrt{3}$

Solution

Note that each of the new equilateral triangles has side length $9 \div 3 = 3$. Hence each has area

$$\frac{3^2\sqrt{3}}{4} = \frac{9\sqrt{3}}{4}.$$

The total area of the second figure is then

$$\frac{81}{4}\sqrt{3} + 3 \times \frac{9\sqrt{3}}{4} = \frac{108\sqrt{3}}{4} = 27\sqrt{3}.$$

(c) **What is the area of the third figure?**

Answer

$30\sqrt{3}$

Solution

Note that each of the new equilateral triangles has side length $3 \div 3 = 1$. Hence each has area

$$\frac{1^2\sqrt{3}}{4} = \frac{1\sqrt{3}}{4}.$$

This time there are a total of

$$2 \times 6 = 12$$

new triangles. The total area of the third figure is thus

$$27\sqrt{3} + 12 \times \frac{1\sqrt{3}}{4} = 30\sqrt{3}.$$

Problem 8.23 (2003 AMC 8) **In the figure, the area of square** $WXYZ$ **is 25 cm^2. The four smaller squares have sides 1 cm long, either parallel to or coinciding with the sides of the large square. In** $\triangle ABC$, $AB = AC$, **and when** $\triangle ABC$ **is folded over side** $\overline{BC}$, **point** A **coincides with** O, **the center of square** $WXYZ$. **What is the area of** $\triangle ABC$, **in square centimeters?**

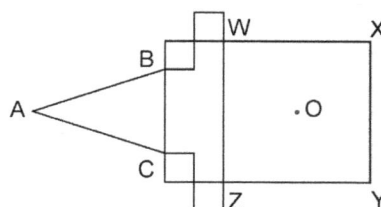

Answer

$$\frac{27}{4}$$

Solution

Given that the area of square WXYZ is 25 cm^2, the side length of square WXYZ is 5 cm.

To determine the height of triangle ABC, note that the height is equal to the length determined by point O and the line segment BC. Since point O is the center of the square, the distance between point O and line segment WZ is 2.5.

The length of line segment WZ and BC is 2. Therefore, the height of the triangle is 4.5. The base length of the triangle is 3 since the side length of the square is 5 which is composed of the length of the base of the triangle plus two unit squares.

Therefore the area of the triangle is

$$\frac{1}{2}(4.5)(3) = \frac{27}{4}.$$

Problem 8.24 **In parallelogram** $ABCD$ **as shown,** $BC = 10$. **Triangle** BCE **is a right triangle where** $\overline{BE}$ **is the hypotenuse, and** $EC = 8$. **Given that** $[ABG] + [CDF] - [EFG] = 10$, **find the length of** $\overline{CF}$.

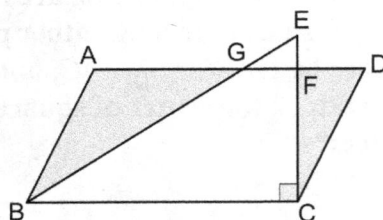

Answer

5

Solution

Note that

$$[ABCD] - [EBC] = [ABG] + [CDF] + [BCFG] - [EFG] - [BCFG] = 10.$$

Therefore,

$$BC \times CF - \frac{1}{2} \times BC \times CE = 10.$$

So,

$$10 \times CF - \frac{1}{2} \times 10 \times 8 = 10.$$

Solving for CF yields $CF = 5$.

Problem 8.25 **In the diagram,** $\triangle ABC$ **and** $\triangle DEF$ **are two congruent isosceles right triangles. Given that** $ADFC$ **is a** 4×3 **rectangle, find the area of the shaded region.**

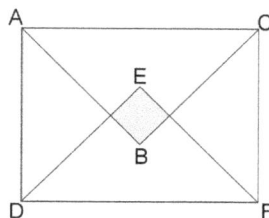

Answer

$\dfrac{1}{2}$

Solution 1

Let G and H be the intersections of the pairs $\overline{AB}, \overline{DE}$ and $\overline{BC}, \overline{EF}$ respectively.

Note that $\triangle AGD, \triangle CHF$ are also isosceles right triangles (congruent to each other, but not $\triangle ABC, \triangle DEF$).

Therefore $GEHB$ is a square, with side length

$$AB - AG = \frac{4}{\sqrt{2}} - \frac{3}{\sqrt{2}} = \frac{1}{\sqrt{2}}.$$

Thus, the shaded region has area $\left(\frac{1}{\sqrt{2}}\right)^2 = \frac{1}{2}$.

Solution 2

Recall that the diagonals of a square are perpendicular and divide the square into 4 congruent isosceles right triangles. Hence, if $ACFD$ were a square, B, E would be the same point.

Therefore, if $\triangle DEF$ is shifted up one unit (resulting in the diagram given), the length of $BE = 1$. As the shaded region is thus a square with diagonal 1, it has area $\frac{1}{2}$.

9. Circle Around

Problem 9.1 **What is the area of a circle with radius 5? Circumference?**

Answer

25π; 10π

Solution

The area of the circle is
$$(5^2)\pi = 25\pi.$$

The circumference of the circle is

$$(2 \times 5)\pi = 10\pi.$$

Problem 9.2 **In the following diagram, circle O has a radius length of 3 cm and a ring with thickness measure 1 cm is placed around circle O 1 cm apart. Is the area of circle O greater than the area of the ring?**

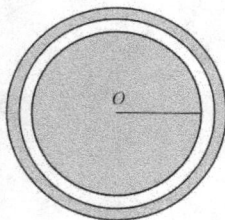

Answer

No

Solution

The area of circle O is

$$3^2\pi = 9\pi$$

and the area of the ring is

$$(5^2 - 4^2)\pi = 9\pi.$$

Problem 9.3 **Find the circumference of a circle with area 16π.**

Answer

8π

Solution

A circle with area 16π has radius 4. Therefore, the circumference of the circle is

$$(2 \times 4)\pi = 8\pi.$$

Problem 9.4 **Find the area of a circle with circumference 16π.**

Answer

64π

Solution

A circle with circumference 16π has radius 8. Therefore, the area of the circle is
$$8^2\pi = 64\pi.$$

Problem 9.5 **Tom and Jerry each eat some pizza. Tom eats a quarter of his pizza, which has a radius of 10 inches. Jerry eats all of his pizza, which has a diameter of 10 inches.**

(a) **Who has eaten more crust?**

Answer

Jerry

Solution

Jerry eats the circumference of his pizza in crust: $2\pi 5 = 10\pi$. Tom each $1/4$ of the circumference of his pizza in crust:

$$\frac{1}{4} \times 2\pi \times 10 = \frac{1}{4} \times 20\pi = 5\pi.$$

(b) **Who has eaten more pizza in total?**

Answer

Both have eaten the same

Solution

Jerry eats the area of his pizza:

$$5^2\pi = 25\pi.$$

Tom each $1/4$ of the area of his pizza:

$$\frac{1}{4} \times \pi \times 10^2 = \frac{1}{4} \times 100\pi = 25\pi.$$

Problem 9.6 **Given a semicircle, let $\overline{AB}$ be its diameter, and O be the center. Let the radius be 4. Randomly select a point C on the arc. What is the maximum possible area of $\triangle ABC$?**

Answer

16.

Solution

The maximum area occurs when C is at the midpoint of arc $\overparen{AB}$. At this point, the area of triangle ABC is 16.

Problem 9.7 **Find the areas of the shaded regions in each of the diagrams. Note: for the most part shapes are drawn to scale, so you may assume angles that look right are in fact 90°, etc.**

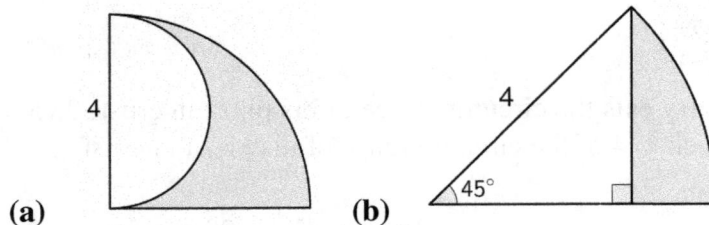

(a)   (b)

Answer

(a) 2π, (b) $2\pi - 4$.

Solution

For (a), we have a quarter circle of radius 4 minus a semicircle of radius 2:

$$\frac{1}{4}\pi \times 4^2 - \frac{1}{2}\pi \times 2^2 = 2\pi.$$

For (b), we have

$$45 \div 360 = \frac{1}{8}$$

of a circle of radius 4 minus a right isosceles triangle with hypotenuse 4:

$$\frac{1}{8}\pi \times 4^2 - \frac{1}{4} \times 4^2 = 2\pi - 4.$$

Problem 9.8 **Find the areas of the shaded regions in each of the diagrams.**

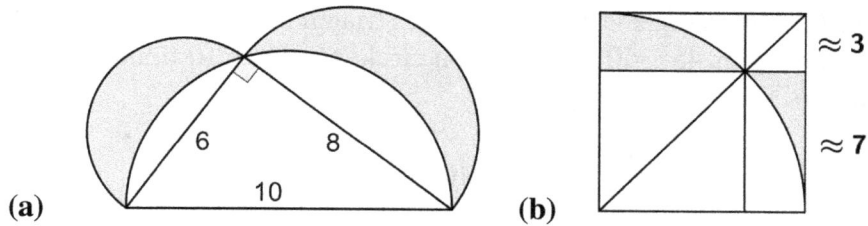

(a)   **(b)**

(a) 24, (b) ≈ 21.

Solution

For (a), we have a semicircle of radius 3 plus a semicircle of radius 4 plus a right triangle with legs 6, 8 minus a semicircle of radius 5:

$$\frac{1}{2}\pi \times 3^2 + \frac{1}{2}\pi \times 4^2 + \frac{1}{2} \times 6 \times 8 - \frac{1}{2}\pi \times 5^2 = 24.$$

(Note: we have shown that the two shaded regions has the same area as the triangle!)

For (b), the two shaded regions can be combined to form a rectangle with approximate dimensions 3 by 7, so the area is approximately 21.

Problem 9.9 **Find the areas of the shaded regions in each of the diagrams.**

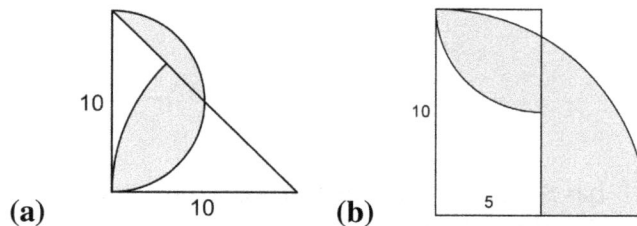

(a)   **(b)**

Answer

(a) $25\pi - 50$, (b) $\dfrac{125\pi}{4} - 50$.

Solution

For (a), the triangle is a right isosceles triangle, so we have a semicircle of radius 5 plus $45/360 = 1/8$th of a circle of radius 10 minus an isosceles right triangle with legs 10:

$$\frac{1}{2}\pi \times 5^2 + \frac{1}{8}\pi \times 10^2 - \frac{1}{2} \times 10^2 = 25\pi - 50.$$

For (b), we have a quartercircle of radius 10 plus a quartercircle of radius 5 minus a rectangle with base and height $5, 10$:

$$\frac{1}{4}\pi \times 10^2 + \frac{1}{4}\pi \times 5^2 - 5 \times 10 = \frac{125\pi}{4} - 50.$$

Problem 9.10 **Find the areas of the shaded regions in each of the diagrams.**

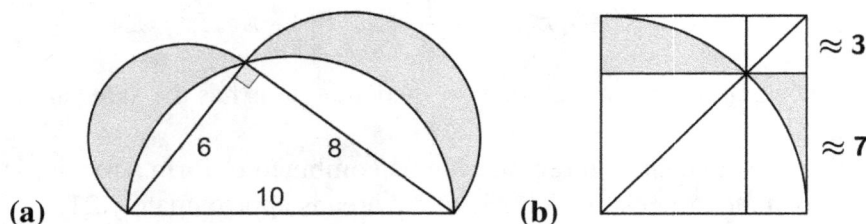

(a)   (b)

Answer

(a) 24, (b) ≈ 21.

Solution

For (a), we have a semicircle of radius 3 plus a semicircle of radius 4 plus a right triangle with legs $6, 8$ minus a semicircle of radius 5:

$$\frac{1}{2}\pi \times 3^2 + \frac{1}{2}\pi \times 4^2 + \frac{1}{2} \times 6 \times 8 - \frac{1}{2}\pi \times 5^2 = 24.$$

(Note: we have shown that the two shaded regions has the same area as the triangle!)

For (b), the two shaded regions can be combined to form a rectangle with approximate dimensions 3 by 7, so the area is approximately 21.

Problem 9.11 **In the diagram below, the area of the shaded region is 4. What is the area of the semicircle with diameter $\overline{OA}$?**

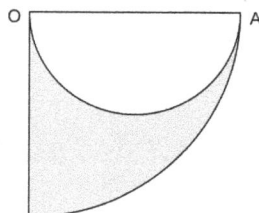

Answer

4.

Solution

The area of the quartercircle is $\frac{OA^2\pi}{4}$, and the area of the semicircle is

$$\frac{(OA/2)^2\pi}{2} = \frac{OA^2\pi}{8},$$

therefore the area of the semicircle is half of that of the quartercircle. Thus the area of the specified semicircle is 4.

Problem 9.12 **Suppose you have a triangle and a semicircle as in the diagram below. Find the difference between the area of region A and region B.**

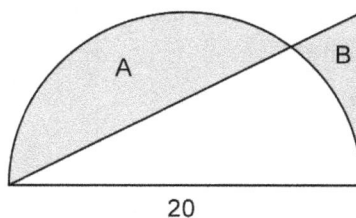

Answer

10.

Solution

The area A minus the area B is equal to the semicircle minus the triangle:

$$\frac{10^2\pi}{2} - \frac{20 \times 10}{2} = 50\pi - 100.$$

Problem 9.13 In the diagram, $ABCD$ is a parallelogram, O is the center of the circle. Given that $[ABCD] = 8$, find the area of the shaded region $\triangle BOC$.

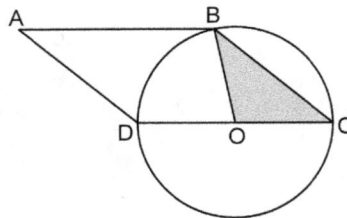

Answer

2.

Solution

Since $\overline{CD}$ is a diameter,

$$[BOC] = [ABCD] \div 4 = 2.$$

Problem 9.14 All the smaller circles in the diagram below have radii 1. Find the perimeter of the shaded region.

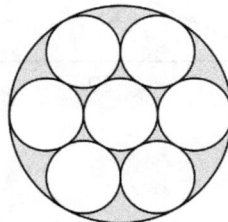

Answer

20π.

Solution

The big circle has radius 3. The perimeter of the shaded region equals the sum of the perimeters of all the circles. Therefore the answer is

$$7 \times (2\pi) + (2 \times 3)\pi = 20\pi.$$

Problem 9.15 **(2010 AMC 8) A decorative window is made up of a rectangle with semicircles at either end as in the diagram below.**

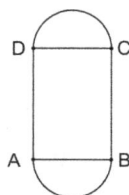

The ratio of AD to AB is $3 : 2$. And AB is 30 inches. What is the ratio of the area of the rectangle to the combined area of the semicircles?

Answer

$6 : \pi$

Solution

Using the ratio given in the problem, we can set up a proportion to determine the length of AD:

$$\frac{AD}{AB} = \frac{AD}{30} = \frac{3}{2}$$

Solving for AD yields 45. We continue with the problem by calculating the combined area of semicircle. This can be achieved by putting together semicircle AB and CD to get a circle with radius 15. Thus, the area is

$$\pi(15^2) = 225\pi.$$

The area of the rectangle is

$$30 \times 45 = 1350$$

and the ratio is:

$$\frac{1350}{225\pi} = \frac{6}{\pi}$$

The answer is $6 : \pi$.

Problem 9.16 **Find the perimeter of the shaded regions in each of the diagrams. Note: for the most part shapes are drawn to scale, so you may assume angles that look right are in fact $90°$, etc.**

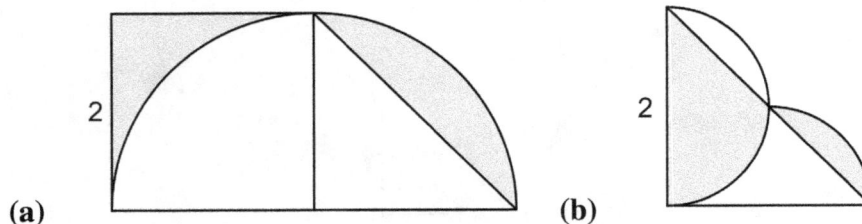

(a)   (b)

Answer

(a) $4 + 2\sqrt{2} + 2\pi$. (b) $\pi + 2\sqrt{2}$

Solution

For (a), the perimeter consists of two sides of a square with side length 2, the hypotenuse of a 45-45-90 triangle with leg length 2, and half the circumference of the circle with radius 2. Therefore, the perimeter of the shaded figure is

$$4 + 2\sqrt{2} + 2\pi.$$

For (b), the perimeter of the shaded region consists of a 45-45-90 triangle and half the circumference of a circle with diameter 2. Thus, the perimeter of the shaded region is

$$\pi + 2\sqrt{2}.$$

Problem 9.17 **Suppose you have a circle of radius 1. A rectangle is inscribed in the circle, and a rhombus is inscribed in the rectangle, as shown in the diagram. What is the side length of the rhombus?**

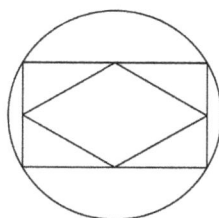

Answer

1.

Solution

The side length of the rhombus is the same as the radius of the circle. So the answer is 1.

Problem 9.18 In the diagram below, the largest circle has radius 5, and the two smaller circles have radii 3 and 4 respectively. The region A is the overlapped region of the two smaller circles. Find the difference between the area of the shaded region and the area of region A.

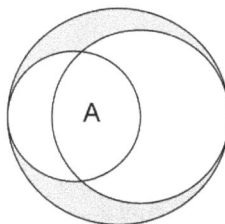

Answer

0.

Solution

Since $3^2 + 4^2 = 5^2$, the area of the largest circle equals the sum of the two smaller circles. Thus the region A and the shaded region have the same area. Therefore the answer is 0.

Problem 9.19 In the diagram, the area of region A equals the area of region B plus $50\pi - 100$. **Find the height of the triangle.**

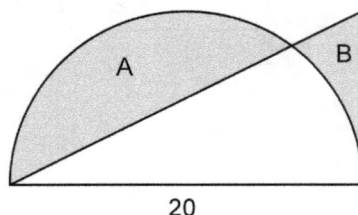

Answer

   10

Solution

Let the height be x. The area A minus the area B is equal to the semicircle minus the triangle:

$$\frac{10^2\pi}{2} - \frac{20x}{2} = 50\pi - 100,$$

Thus

$$50\pi - 10x = 50\pi - 100$$

so we see that x must be 10.

Problem 9.20 (2013 AMC 8) A 1×2 rectangle is inscribed in a semicircle with **the longer side on the diameter. What is the area of the semicircle?**

Answer

   π

Solution

Since semicircles are symmetric, the center is exactly at the midpoint of the side on the rectangle. Therefore, by the Pythagorean theorem, the radius has length

$$\sqrt{1^2 + 1^2} = \sqrt{2}$$

and the area is

$$\frac{2\pi}{2} = \pi.$$

Problem 9.21 Congruent circles with centers A and B intersect such that AB is a radius of each circle. If $AB = 3$, what is the area of the intersecting region shared by the two circles?

Answer

$$6\pi - \frac{9}{2}\sqrt{3}$$

Solution

Let us determine the area of the sector of the circle with center A that covers most of the region of interest. This sector is $\frac{1}{3}$ of the entire circle, so the area of this sector is

$$\frac{1}{3}\pi \times 3^2 = 3\pi.$$

Subtract the area of the triangle formed by connecting the points of intersection of the circles. Using special right triangles, we observe that the area of the triangle is

$$\frac{1}{2} \times \frac{3}{2} \times 3\sqrt{3} = \frac{9}{4}\sqrt{3}.$$

Therefore, the area of the intersecting region is

$$2 \times (3\pi - \frac{9}{4}\sqrt{3}) = 6\pi - \frac{9}{2}\sqrt{3}.$$

Problem 9.22 Three circles having each a radius of 3 cm are tangent to each other. Consider the triangle formed by joining their three centers. What is the area of the region inside this triangle but outside the three circles.

Answer

$$9\sqrt{3} - \frac{9}{2}\pi$$

Solution

The area is equal to the area of the triangle minus the area of the three sectors of each circle inside the triangle. The area of the triangle (using $30-60-90$ special right triangles to determine the height) is

$$\frac{1}{2} \times 6 \times 3\sqrt{3} = 9\sqrt{3}$$

The total area of all three sectors of the circle is equivalent to the area of the semicircle of radius 3. Thus, the area that the circles contribute is equal to

$$\frac{1}{2}\pi \times 3^2 = \frac{9}{2}\pi.$$

The answer is $9\sqrt{3} - \frac{9}{2}\pi$.

Problem 9.23 Let $ABCD$ **be a square with edge length of** 3 **cm. We draw a circle of center** A **and radius** 3 **cm and a circle of center** C **and radius** 3 **cm. What is the ratio of the area of the circle's intersection to the area of the square?**

Answer

$$\left(\frac{1}{2}\pi - 1\right) : 1$$

Solution

Note that the area of the intersection of the circles inside the square can be determined by finding the area of two quarter circles and subtracting the area of the square. Therefore, the area of the intersection of the circles is

$$\frac{1}{4}\pi \times 3^2 + \frac{1}{4}\pi \times 3^2 - 3^2 = 9 \times \left(\frac{1}{2}\pi - 1\right).$$

The ratio of the area of the circle's intersection to that area of the square is $\left(\frac{1}{2}\pi - 1\right) : 1$.

Problem 9.24 **(2010 AMC 8) The two circles pictured have the same center** C**.**

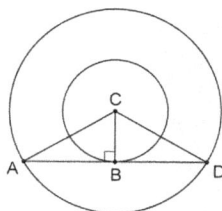

Chord $\overline{AD}$ is tangent to the inner circle at B, AC is 10, and chord $\overline{AD}$ has length 16. What is the area between the two circles?

64π

Solution

Since we are given that $\triangle ACD$ is isosceles, CB bisects AD. Thus, $AB = BD = 8$. From the Pythagorean Theorem,

$$CB = \sqrt{10^2 - 8^2} = 6$$

and the area between the two circles is

$$100\pi - 36\pi = 64\pi.$$

Problem 9.25 Let ABC be an equilateral triangle with coordinates $A = (0,0)$ and $B = (1,0)$. Triangle ABC is rotated counterclockwise about point A so that the image of C lies on the $y-$axis. Find the length of the path formed by point C during the rotation.

Answer

$$\frac{\pi}{6}$$

Solution

Recall that an equilateral triangle have $60°$ angles. By rotating the triangle so that the image of C lies on the $y-$axis, we require a counter-clockwise rotation of 30° to achieve this. Since this was a rotation of 30°, the path of point C is a 30° arc of a circle with radius 1. Therefore, the length of the path of point C is

$$\frac{30}{360} \times 2\pi \times 1 = \frac{\pi}{6}.$$

10. Circle Back

Problem 10.1 Points A and B lie on a circle centered at point O with radius length 9 in. Suppose that the measure of $\angle AOB$ is $120°$. **What is the length of arc AB?**

Answer

6π in

Solution

Given that the radius of the circle is 9 in., the circumference of the circle is
$$(2 \times 9)\pi = 18\pi.$$
Note that $\angle AOB$ creates a sector that is $\frac{1}{3}$ of the entire circle. Therefore, the length of arc AB is $\frac{1}{3}$ of the circumference, or 6π in.

Problem 10.2 Points A and B lie on a circle centered at point O with radius length 10 in. Suppose that the area measure of the sector formed by $\angle AOB$ is 20π in^2. **What is $\angle AOB$?**

Answer

72°

Solution

Given that the radius of the circle is 10 in., the area of the circle is $10^2\pi$ in^2.
Since 20π is $\frac{1}{5}$ of 100π,

$$\angle AOB = \frac{1}{5} \times 360° = 72°.$$

Problem 10.3 Find the areas of the shaded regions in each of the diagrams below. **Note: for the most part shapes are drawn to scale, so you may assume angles that look right are in fact** $90°$**, etc.**

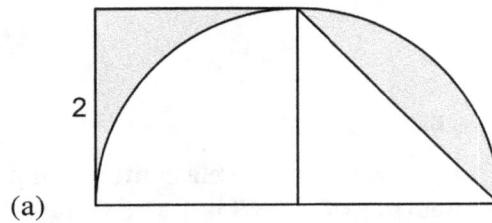

(a)

Answer

2

Solution

Note the two shaded regions combine to form half a 2 by 2 square. Thus the area is
$$\frac{1}{2} \times 2 \times 2 = 2.$$

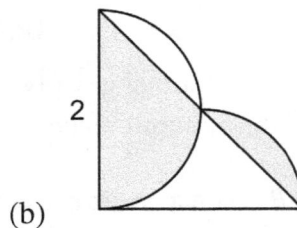

(b)

Answer

$$\frac{\pi}{2}$$

Solution

Note the two shaded regions combine to form a semicircle with radius 1. Thus the area is

$$\frac{1}{2}\pi \times 1^2 = \frac{\pi}{2}.$$

Problem 10.4 **A circular dining table has diameter 2 meters and height 1 meter. A square tablecloth is placed on the table, and the four corners of the tablecloth just touch the floor. Find the area of the tablecloth in square meters.**

Answer

8m^2

Solution

As shown in the diagram, the tablecloth's diagonal is 4 meters, thus the area is

$$4^2 \div 2 = 8$$

m^2.

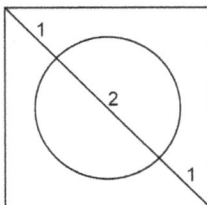

Problem 10.5 **The shape in the diagram consists of three semicircles with radii 5, 3, and 2, arranged as shown. Calculate the perimeter and area of the shaded region.**

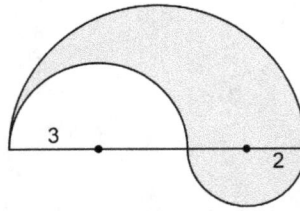

Answer

$10\pi, 10\pi$

Solution

The perimeter of the shape in the above diagram is

$$2\pi \times \frac{3}{2} + 2\pi \times \frac{2}{2} + 2\pi \times \frac{5}{2} = 10\pi,$$

and area of the shape in the above diagram is

$$\frac{5^2\pi}{2} - \frac{3^2\pi}{2} + \frac{2^2\pi}{2} = 10\pi.$$

Problem 10.6 In the diagram below, the 4 smaller circles are all congruent and pass through a common point, and are all internally tangent to the bigger circle. Find the ratio between the perimeters of the shaded region and the bigger circle.

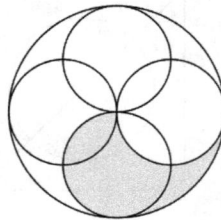

Answer

$3 : 4.$

Solution

The perimeter of the shaded region is the sum of two semicircles with the smaller radius and a quartercircle with the larger radius.

Since the smaller radius is half of the larger radius, the ratio in question is $3:4$.

Problem 10.7 A sheep is tied with a rope to the upper-left corner of the square barn on the grass field. The length and width of the barn are 10, as shown in the diagram, and the length of the rope is 20. Find the area of the region that the sheep can reach.

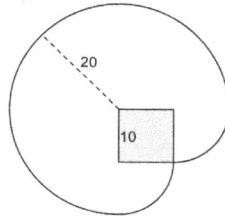

Answer

350π

Solution

The region can be divided into a sector of $270°$ of the circle with radius 20 and two quartercircles with radii 10. Thus the area is

$$\frac{3}{4} \times 20^2 \pi + 2 \times \frac{1}{4} \times 10^2 \pi = 350\pi.$$

Problem 10.8 In the diagram, the radius of the circle is 4. If we are also given that the area of the shaded region is 14π, find the area of the triangle.

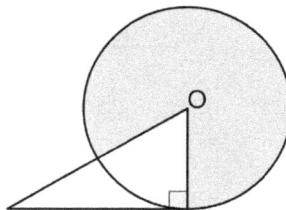

Note: This diagram does not reflect the actual shape: the angle of the unshaded sector is *not* 60°.

Answer

8

Solution

The area of the full circle is

$$4^2\pi = 16\pi.$$

Therefore, the shaded region is

$$\frac{14}{16} = \frac{7}{8}$$

of the full circle, so the angle at point O is

$$\frac{360°}{8} = 45°.$$

So the triangle is an isosceles right triangle, and its area is

$$4^2 \div 2 = 8.$$

Problem 10.9 Six cylindrical pencils are tied together with a rubber band. A cross section of it is shown in the diagram. The radius of the pencils is 1. Find the current length of the rubber band.

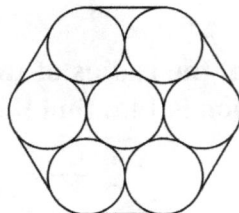

Answer

$2\pi + 12.$

Solution

The rubber band consists of 6 arcs of 60° each, and 6 line segments of length 2 each. Thus the total length is $2\pi + 12$.

Problem 10.10 Equilateral triangle ABC pictured below has side length 1. It is placed on a straight line at position I as shown. Roll the triangle along the line about vertex C to reach position II, and roll again to reach position III. Find the total length of the path that vertex A traveled.

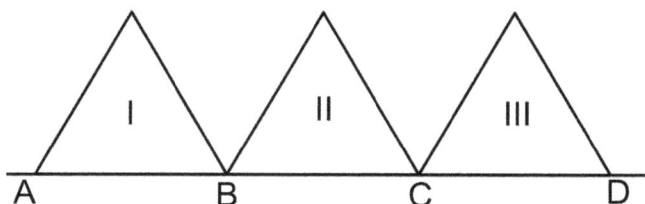

Answer

$$\frac{4\pi}{3}$$

Solution

From I to II, vertex A traveled along an arc of 120°, which is 1/3 of a full circle. From II to III, vertex A traveled along another arc of 120°. So the total length is

$$\frac{2\pi}{3} \times 2 = \frac{4\pi}{3}.$$

Problem 10.11 (**2010 AMC 8**) Semicircles POQ and ROS pass through the center O.

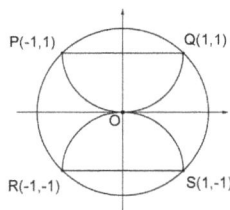

What is the ratio of the combined areas of the two semicircles to the area of circle
O?

Answer

$$\frac{1}{2}$$

Solution

By the Pythagorean Theorem, the radius of the larger circle is

$$1^2 + 1^2 = \sqrt{2}.$$

Therefore, the area of the larger circle is

$$(\sqrt{2})^2 \pi = 2\pi.$$

Using the coordinates given in the problem, we find that the radius of the two semicircles to be 1 and the area of the two semicircles is

$$1^2 \pi = \pi.$$

Therefore, the ratio of the combined areas of the two semicircles to the area of circle O is $\frac{1}{2}$.

Problem 10.12 **In the diagram below, the area of the shaded region (area between the two squares) is 200. What is the area of the annular region (between the two circles)?**

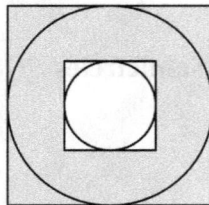

Answer

50π

Solution

The side length a of the bigger square equals the diameter of the bigger circle, and the side length b of the smaller square equals the diameter of the smaller circle. Thus

$$a^2 - b^2 = 200.$$

The area of the annular region is

$$\left(\frac{a}{2}\right)^2 \pi - \left(\frac{b}{2}\right)^2 \pi = \frac{(a^2 - b^2)\pi}{4} = 50\pi.$$

Problem 10.13 Given two circles, whose perimeters have ratio $3 : 2$. Also given that the difference between their areas is 10 cm^2. What is the sum of their areas (in cm^2)?

Answer

26 cm^2

Solution

If the perimeters have ratio $3 : 2$, the areas must have ratio $9 : 4$. Let the areas be $9x$ and $4x$ respectively, then

$$9x - 4x = 10,$$

so $x = 2$. Therefore the answer is $9x + 4x = 26$ cm^2.

Problem 10.14 Rectangle I has length 15 and width 8, and is placed on a straight line as shown in the diagram. Roll the rectangle about vertex B by $90°$ to reach position II. Then roll about vertex C, and so on, until the original vertex A reaches point E. Calculate the total length of the path that vertex A traveled.

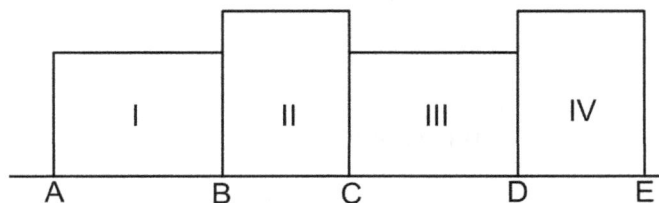

Answer

20π.

Solution

Label the upper-left corner of rectangle II as A_1, and the upper-right corner of rectangle III as A_2.

From A to A_1, it's a quartercircle of radius 15. From A_1 to A_2, it's a quartercircle of radius

$$\sqrt{8^2 + 15^2} = 17.$$

From A_2 to E, it's a quartercircle of radius 8. So the length of the total path is

$$\frac{2(8 + 15 + 17)\pi}{4} = 20\pi.$$

Problem 10.15 (2008 AMC 8) Margie's winning art design is shown.

The smallest circle has radius 2 inches, with each successive circle's radius increasing by 2 inches. Approximately what percent of the design is black?

Answer

$\approx 42\%$

Solution

The area of the entire art design is

$$(12^2)\pi = 144\pi.$$

The area of the black region is determined by

$$(10^2 - 8^2)\pi + (6^2 - 4^2)\pi + 2^2\pi = 60\pi.$$

The percentage of the design that is black is

$$\frac{60\pi}{144\pi} = \frac{5}{12} \approx 42.$$

Problem 10.16 **Find the perimeter of the shaded regions in each of the diagrams. Note: for the most part shapes are drawn to scale, so you may assume angles that look right are in fact 90°, etc.**

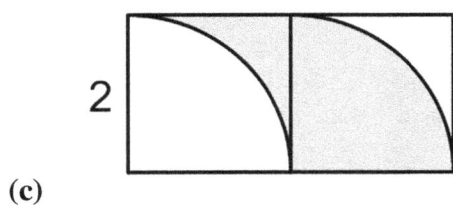

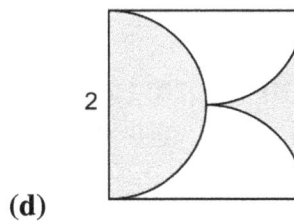

(c)   (d)

Answer

(a) $4 + 2\pi$, (b) $4 + 2\pi$.

Solution

For (a), note the perimeter of the shaded region represents the length of two sides of a square with side length 2 and half the perimeter of a circle with diameter 2. Therefore, the perimeter is $4 + 2\pi$.
For (b), note the perimeter of the shaded regions represents the circumference of the circle with diameter 2 plus two side lengths of a square with radius 2. Thus the perimeter is $4 + 2\pi$.

Problem 10.17 **Find the perimeter of the shaded regions in each of the diagrams.**

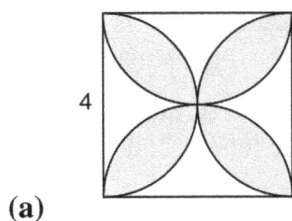

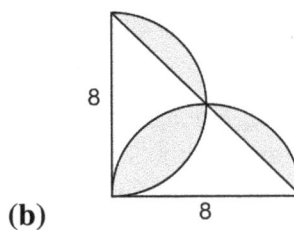

(a)   (b)

Answer

(a) 8π, (b) 8π.

Solution

For (a), note that the perimeter of the shaded region is equal to 4 halves of the circumference of the circle with diameter 2. Therefore, the perimeter is $2\pi \times 4 = 8\pi$.

For (b), we use a similar idea to (a). Note that the perimeter of the shaded region is equal to the circumference of a circle with diameter 8. Therefore, the answer is 8π.

Problem 10.18 (2011 AMC 8) A circle with radius 1 is inscribed in a square and circumscribed about another square as shown.

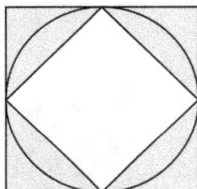

What is the ratio of the circle's shaded area to the area between the two squares?

Answer

$$\frac{\pi - 2}{2}$$

Solution

Firstly, note that the side length of the square is equal to the diameter of the circle. Therefore, the side length of the square is 2.

Temporarily ignoring the circle, we observe that the shaded region represents $\frac{1}{2}$ the area of the square with side length 2. Therefore, the area of the shaded region formed by the squares is 2.

Now, let us consider the circle. Note that the diagonals of the smaller square corresponds to the diameter of the circle. We had previously established that the area of the smaller square is 2. The area of the circle with radius 1 is

$$1^2\pi = \pi.$$

Therefore, the area of the shaded region composed of the smaller square and the circle is $\pi - 2$. The desired ratio is then $\frac{\pi-2}{2}$.

Problem 10.19 Congruent circles centered at O and P are externally tangent, and a line through P is tangent to O at A. If $AP = 15$, what is the length of the radius of circle P?

Answer

$5\sqrt{3}$

Solution

Let M be the point where the two circles are tangent to each other and let x be the radius of the circle. Then,

$$MP = OM = OA = x.$$

Furthermore, OA is perpendicular to AP, forming a right triangle. Applying Pythagorean Theorem, we have,

$$x^2 + 15^2 = (2x)^2$$

or $x = 5\sqrt{3}$.

Problem 10.20 Let A be a circle with radius 1. Let B be the inscribed regular hexagon inside circle A. Let C be the inscribed circle inside hexagon B. What is the ratio of the area of circle C to the area of circle A?

Answer

$3 : 4$

Solution

The area of circle A is simply $1^2\pi = \pi$. Since regular hexagons can be decomposed into six equilateral triangles, the side length of hexagon B is 1.

Therefore, the radius of circle C is equal to the height of the equilateral triangle with side length 1.

That is, radius of circle C is $\frac{\sqrt{3}}{2}$. Therefore, the ratio of the area of circle C to the area of circle A is $\pi(\frac{\sqrt{3}}{2})^2 : \pi$ or $3 : 4$.

Problem 10.21 Three congruent circles with centers A, B, and C intersect such that AB, AC, and BC are radii of two circles. If $AB = 3$, what is the area of the intersecting region shared by the three circles?

Answer

$\frac{9}{2}(\pi - \sqrt{3})$

Solution

Note that the area of the region of interest can be determined by taking the area of three circular sectors and subtracting it by the area of two equilateral triangles. The area of the sector is

$$\frac{1}{6}\pi \times 3^2 = \frac{3}{2}\pi$$

and the area of the equilateral triangle with side length 3 is

$$\frac{1}{2}3 \times \frac{3\sqrt{3}}{2} = \frac{9\sqrt{3}}{4}.$$

Therefore,

$$3 \times \frac{3}{2}\pi - 2 \times \frac{9\sqrt{3}}{4} = \frac{9}{2}(\pi - \sqrt{3}).$$

Problem 10.22 Rectangle I has length 4 and width 3, and is placed on a straight line as shown in the diagram. Roll the rectangle about vertex B by $90°$ to reach position II. Then roll about vertex C, and so on, until the original vertex A reaches point E. Calculate the total length of the path that vertex A traveled.

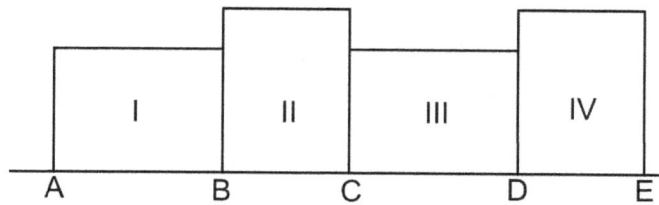

Answer

6π

Solution

Label the upper-left corner of rectangle II as A_1, and the upper-right corner of rectangle III as A_2.

From A to A_1, it's a quartercircle of radius 4. From A_1 to A_2, it's a quarter-circle of radius

$$\sqrt{3^2 + 4^2} = 5.$$

From A_2 to E, it's a quartercircle of radius 3. So the length of the total path is

$$\frac{2 \times (4+5+3)\pi}{4} = 6\pi.$$

Problem 10.23 **Using the above problem, find the area of the region formed by the path of vertex A.**

Answer

$\frac{25}{2}\pi + 12$

Solution

Note that the regions formed by the path traveled by vertex A consists of one quarter of a circle with radius 3, one quarter of a circle with radius 4, one quarter of a circle with radius 5, and two triangles with base length 4 and height 3. Therefore, the answer is

$$\pi \times \frac{3^2 + 4^2 + 5^2}{4} + 12 = \frac{25}{2}\pi + 12.$$

Problem 10.24 Given a semicircle with diameter $\overline{AB}$, where $AB = 3$. Rotate this semicircle about point A by $60°$ counterclockwise, so that the point B reaches B', as shown in the diagram. Find the area of the shaded region.

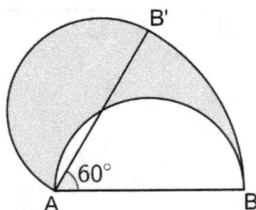

Answer

$$\frac{3\pi}{2}$$

Solution

The area of the shaded region is the semicircle plus the sector of $60°$, minus the semicircle, thus it is just the sector (radius 3). So the answer is

$$\frac{1}{6} \times 3^2 \pi = \frac{3\pi}{2}.$$

Problem 10.25 Using the same figure above, find the perimeter of the shaded region.

Answer

$$4\pi$$

Solution

The perimeter of the shaded region is the perimeter of two semicircles plus the perimeter of the sector of $60°$. The answer is

$$\frac{3}{2}\pi + \frac{3}{2}\pi + \frac{1}{6} \times 6\pi = 4\pi.$$

11. A Whole New Dimension

Problem 11.1 Recall that in a cube such as the one in the diagram below,

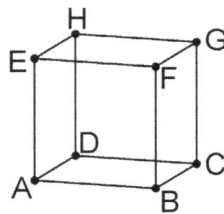

there are a total of 12 edges. Consider $\overline{AB}$. Which of the other edges intersect $\overline{AB}$? Which are parallel to $\overline{AB}$? Which are skew to $\overline{AB}$?

Answer

Intersect: $\overline{AD}, \overline{AE}, \overline{BC}, \overline{BF}$ Parallel: $\overline{CD}, \overline{EF}, \overline{GH}$ Skew: $\overline{CG}, \overline{DH}, \overline{EH}, \overline{FG}$

Problem 11.2 Suppose you have a box (rectangular prism) that is 2 feet long, 1 foot wide, and has a height of 6 inches. (1 foot equals 12 inches)

(a) **How much space is inside the box? That is, what is the volume of the box?**

> **Answer**

> 1 ft^3 or 1728 in^3

> **Solution**

> In feet, the box has dimensions $2 \times 1 \times .5$ so it has volume 1 ft^3 (cubic feet). In inches, it is $24 \times 12 \times 6 = 1728$ in^3.

(b) **What is the surface area of the box?**

> **Answer**

> 7 ft^2 or 1008 in^2

> **Solution**

> The surface area is $2(2 \times 1 + 2 \times 0.5 + 1 \times .5) = 7$ ft^2 (square feet) or, $2(24 \times 12 + 24 \times 6 + 12 \times 6) = 1008$ in^2.

Problem 11.3 Suppose we have a sphere with radius 6.

(a) **Find the volume of the sphere.**

> **Answer**

> 228π

> **Solution**

> Using the formula we have the volume is $\dfrac{4\pi}{3} \times 6^3 = 228\pi$.

(b) **Find the surface area of the sphere.**

> **Answer**

> 144π

> **Solution**

> Using the formula we have the surface area is $4\pi \times 6^2 = 144\pi$.

Problem 11.4 Find the volume and surface area of a ball with radius 3.

Answer

$36\pi, 36\pi$

Solution

The volume is $\dfrac{4\pi}{3} \times 3^3 = 36\pi$ and the surface area is $4\pi \times 3^2 = 36\pi$.

Problem 11.5 **Find the volume and surface area of a square pyramid with base length 6 and height 4.**

Answer

$48; 96$

Solution

The volume of a square pyramid can be determined by the following calculation:

$$\frac{1}{3} \times 6 \times 6 \times 4 = 48$$

Note that the height of the triangle can be computed using Pythagorean theorem:

$$\sqrt{3^2 + 4^2} = 5$$

Therefore, the surface area of a square pyramid with the above dimensions is:

$$4 \times \frac{1}{2}(5 \times 6) + 6^2 = 96$$

Problem 11.6 **Compare the volume and surface area of a cube with side length 2 to a rectangular prism with dimensions 1, 2, 4.**

Answer

Same volume but different surface areas

Solution

Note that a cube with side length 2 has volume

$$2^3 = 8$$

and the volume of a rectangular prism is

$$1 \times 2 \times 4 = 8$$

so the volumes are the same. However, the surface area of the cube is

$$6 \times 2^2 = 18$$

and the surface area of the rectangular prism is

$$2(1 \times 2 + 1 \times 4 + 2 \times 4) = 28.$$

Therefore, the surface areas are different.

Problem 11.7 Compare the volume and surface area of a cube with side length 10 to a rectangular prism with dimensions $10, 15, 6$.

Answer

Same surface area but different volume.

Solution

Because surface area of the cube with side length 10 calculated as follows:

$$2(10 \times 10 + 10 \times 10 + 10 \times 10) = 600$$

and the surface area of the rectangular prism is:

$$2(10 \times 15 + 15 \times 6 + 6 \times 10) = 600$$

They have the same surface area. However, the volume of the cube is

$$10 \times 10 \times 10 = 1000$$

and the volume of the rectangular prism is:

$$10 \times 15 \times 6 = 900$$

so the volumes are different.

Problem 11.8 **Find the distance from one corner to the opposite corner (for example the lower front left vertex to the upper back right vertex) of a rectangular prism with dimensions** $3, 4, 12$. **That is, find the length of the line connecting the two corners.**

Answer

13

Solution

First use the Pythagorean theorem to find the distance between opposite corners of the base (i.e. the length of the diagonal of the rectangular base):

$$\sqrt{3^2 + 4^2} = 5.$$

Note then that this diagonal is perpendicular to the vertical edges of the rectangular prism, so we can use the Pythagorean theorem again to find the distance between opposite corners:

$$\sqrt{5^2 + 12^2} = 13.$$

Therefore the length of the line connecting the two corners is 13.

Problem 11.9 **Extending the previous problem, find the distance from one corner to the opposite corner of a** $l \times w \times h$ **rectangular prism. Note this problem can be thought of as an extension of the Pythagorean theorem.**

Answer

$\sqrt{l^2 + w^2 + h^2}$

Solution

Similar to previous questions, we can use the Pythagorean theorem twice the distance is

$$\sqrt{\sqrt{l^2 + w^2}^2 + h^2} = \sqrt{l^2 + w^2 + h^2}$$

as needed.

Problem 11.10 **Suppose a ball has volume $\frac{32\pi}{3}$. Find the surface area of the ball.**

Answer

16π

Solution

Since the ball has volume $\frac{32\pi}{3}$, we see that the radius cubed must be 8, so the radius is 2. Hence the surface area is

$$4\pi \times 2^2 = 16\pi.$$

Problem 11.11 **Compare the volume of a ball with radius 4 to the combined volumes of two balls of radius 3.**

Answer

The one ball of radius 4 has more volume

Solution

The volume of a ball with radius 4 is

$$\frac{4\pi}{3} \times 4^3 = \frac{256\pi}{3}$$

We saw a ball of radius 3 had volume 36π, so two balls have a volume of 72π. Note

$$256 \div 3 \approx 85,$$

so the one ball of radius 4 has a larger volume.

Problem 11.12 **Compare the surface area of a ball with radius 4 to the combined surface areas of two balls of radius 3.**

Answer

The two balls of radius 3 have more combined surface area

Solution

The surface area of a ball with radius 4 is

$$4\pi \times 4^2 = 64\pi.$$

We saw a ball of radius 3 had surface area 36π, so two balls have a surface area of 72π.

Therefore, the two balls of radius 3 have a larger surface area.

Problem 11.13 **How many total pairs of parallel edges does a cube contain?**

Answer

18

Solution

Recall the cube diagram with labeled points.

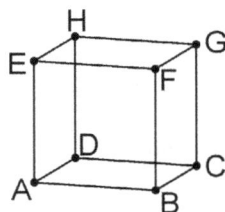

We have that $\overline{AD}, \overline{BC}, \overline{EH}, \overline{FG}$ are all parallel.

This gives

$$3 + 2 + 1 = 6$$

pairs ($\overline{AD}, \overline{BC}$ or $\overline{AD}, \overline{EH}$ or $\overline{AD}, \overline{FG}$ or $\overline{BC}, \overline{EH}$ or $\overline{BC}, \overline{FG}$ or $\overline{EH}, \overline{FG}$.

Similarly, $\overline{AB}, \overline{CD}, \overline{EF}, \overline{GH}$, giving another 6 pairs.

Finally, $\overline{AE}, \overline{BF}, \overline{CG}, \overline{DH}$ are all parallel, giving us 6 pairs again.

Hence there are

$$6 + 6 + 6 = 18$$

total pairs of parallel lines.

Problem 11.14 **Consider a** $8\frac{1}{2} \times 11$ **sheet of paper. One can observe that there are two ways of curling the sheet of paper so that it resembles a cylinder. Find the volumes of all possible cylinders formed by a sheet of** $8\frac{1}{2} \times 11$ **paper.**

Answer

$$\frac{3179}{16\pi}; \frac{2057}{8\pi}$$

Solution

There are two ways to form a cylinder with a sheet of $8\frac{1}{2} \times 11$ paper. If we curl the sheet of paper along the $8\frac{1}{2}$ side, we observe that it forms a cylinder with a circular base with circumference $8\frac{1}{2}$ and height 11. This implies that the radius of the cylinder can be computed by solving the following equation:

$$8\frac{1}{2} = 2\pi r$$

This yields $r = \frac{17}{4\pi}$. Therefore, the volume of this cylinder is:

$$\pi \left(\frac{17}{4\pi}\right)^2 \times 11 = \frac{3179}{16\pi}$$

Alternatively, we can curl the sheet of paper along the 11 side. By doing this, we observe that it forms a cylinder with a circular base with circumference 11 and height $8\frac{1}{2}$. This implies that the radius of the cylinder can be determined by solving the following equation:

$$11 = 2\pi r$$

Solving the above yields $r = \frac{11}{2\pi}$. Therefore, the volume of this cylinder is:

$$\pi \left(\frac{11}{2\pi}\right)^2 \times 8\frac{1}{2} = \frac{2057}{8\pi}$$

Problem 11.15 **It is well known that about** 70% **of the Earth's surface is covered in water. If a sphere of radius** 2 **cm is painted in blue for water and green for land to create a miniature globe, find the area of the surface that is painted blue.**

Answer

5.6π cm^2

Solution

The surface area of a sphere with radius 2 cm is computed as follows:

$$4\pi \times 2 = 8\pi \text{cm}^2$$

If 70% of the sphere's surface is water, then $0.7 \times 8\pi = 5.6\pi$ cm^2 of the surface of the sphere must be painted blue.

Problem 11.16 In the $2 \times 2 \times 2$ cubic figure below, if the path is required to be along the surface of the cube, what is the length of the shortest path from point A to B?

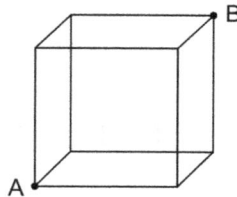

Answer

$2\sqrt{5}$

Solution

The shortest distance satisfying the above conditions can be determined as follows:

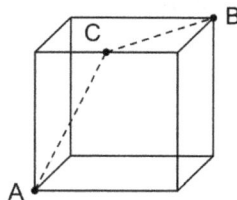

Note that the length of AC and CB can be determined by applying Pythagorean theorem on a right triangle with leg lengths 2 and 1. Therefore,

$$AC = BC = \sqrt{2^2 + 1^2} = \sqrt{5}$$

and the length of the shortest path from point A to B is $2\sqrt{5}$.

Problem 11.17 (2014 AMC 8) A cube with 3-inch edges is to be constructed from 27 smaller cubes with 1-inch edges. Twenty-one of the cubes are colored red and 6 are colored white. If the 3-inch cube is constructed to have the smallest possible white surface area showing, what fraction of the surface area is white?

Answer

$\dfrac{5}{54}$

Solution

The total surface area is

$$6 \times 3^2 = 54$$

square inches.

To hide as many white faces as possible, first note that one of the 6 white cubes should be the middle of the big cube. The other 5 cubes have to have at least one face exposed, but if we put them in the middle of 5 of the faces of the big cube, they will each only have 1 face exposed.

Since each of these exposed faces has area 1 square inch, there is a total of 5 square inches of white area. Hence the fraction of white area to the total surface area is $\frac{5}{54}$.

Problem 11.18 (2013 AMC 8) Isabella uses one-foot cubical blocks to build a rectangular fort that is 12 feet long, 10 feet wide, and 5 feet high. The floor and the four walls are all one foot thick.

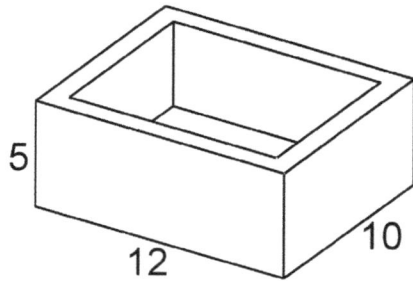

How many blocks does the fort contain?

Answer

280

Solution

Since each block is a one-foot cubical block, we need to find the volume of the fort. Note if the fort was "filled-in" it would have a volume of

$$12 \times 10 \times 5 = 600$$

cubic feet. However, since the walls are all one foot thick, there is actually a box that is $12 - 2 = 10$ feet long, $10 - 2 = 8$ feet wide, and $5 - 1 = 4$ feet high removed from the 600 cubic feet box.

Since the removed box has volume

$$10 \times 8 \times 4 = 320,$$

the fort has volume 280 cubic feet.

Problem 11.19 **Suppose a rectangular block of wood with dimensions 5 cm $\times 6$ cm $\times 5$ cm costs \$50. In dollars, what is the fair price for a rectangular block of the same type of wood with dimensions 15 cm by 30 cm by 40 cm if the price is determined solely by volume?**

Answer

6000

Solution

The volume of the smaller rectangular block is

$$5 \times 6 \times 5 = 150$$

cm^3. The volume of the bigger rectangular block is

$$15 \times 30 \times 40 = 18000$$

cm^3. Since the ratio of the volume of the big block to the smaller block is $120:1$, the fair price for the bigger rectangular block is $50 \times 120 = \$6000$.

Problem 11.20 **An obtuse triangle with dimensions 9, 10, and 17 is rotated about the smallest side so that it creates a three-dimensional solid shown below. Determine the volume of the solid.**

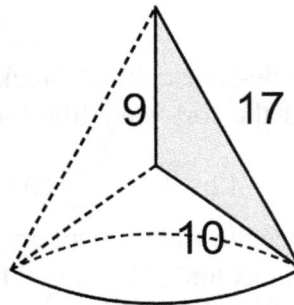

Answer

192π

Solution

Note that in the figure above, there are two cones sharing the same circular base. Let r be the radius of the cones and let h be the height of the smaller cone. Therefore, h and r satisfies

$$h^2 + r^2 = 10^2$$

and

$$(h+9)^2 + r^2 = 17^2.$$

If you recall Pythagorean triples, $h = 6$ and $r = 8$ yields a $6 - 8 - 10$ and $8 - 15 - 17$ Pythagorean triples. Therefore, the volume of the new figure is

$$\frac{1}{3}\pi \times 8^2 \times 15 - \frac{1}{3}\pi \times 8^2 \times 6 = 192\pi.$$

Problem 11.21 **The square pyramid J_1 is a regular square pyramid composed of four equilateral trianglular faces and a square base. Given that the surface area of J_1 is $1 + \sqrt{3}$, find the volume of J_1.**

Answer

$$\frac{\sqrt{2}}{6}$$

Solution

Let s be the side length of J_1. Since J_1 is a figure composed of four equilateral trianglular faces and a square base, the surface area of J_1 in terms of s is

$$s^2 + 4 \times \frac{s}{2} \times \frac{\sqrt{3}s}{2} = (1 + \sqrt{3})s^2.$$

Therefore, since J_1 has surface area $1 + \sqrt{3}$, the side length must be $s = 1$. Note that the height of the equilateral triangle corresponds to the hypotenuse of the right triangle formed by half the side of the square and the height of the square pyramid J_1. This implies that the height of J_1 is equal to

$$\sqrt{\left(\frac{\sqrt{3}}{2}s\right)^2 - \left(\frac{1}{2}s\right)^2} = \frac{1}{\sqrt{2}}s.$$

Therefore, the volume of J_1 is

$$\frac{1}{3} \times s^2 \times \frac{1}{\sqrt{2}}s = \frac{\sqrt{2}}{6}s^3.$$

Since $s = 1$, the volume of J_1 is equal to $\frac{\sqrt{2}}{6}$.

Problem 11.22 A regular square pyramid is placed in a cube so that the base of the pyramid and the base of the cube coincide. The vertex of the pyramid lies on the face of the cube opposite to the base. Suppose that the side length of the cube is 2 in. What is the positive difference of the surface area of the cube and the surface area of the pyramid?

Answer

$20 - 4\sqrt{5}$

Solution

The surface area of the cube is found as:

$$6 \times 2^2 = 24.$$

Note that the height of the triangles that forms the desired pyramid can be determined using Pythagorean theorem on the side length of the cube. Namely, the height of the triangle, h, can be determined as follows:

$$h = \sqrt{2^2 + 1^2} = \sqrt{5}.$$

Therefore, the area of the triangle is

$$\frac{1}{2} \times 2 \times \sqrt{5} = \sqrt{5}.$$

Since a pyramid consists of four of these triangles and the square base, the surface area of the pyramid is $4\sqrt{5} + 4$. Therefore, the positive difference is $24 - (4\sqrt{5} + 4) = 20 - 4\sqrt{5}$.

Problem 11.23 A cube is increased to form a new cube so that the surface area of the new cube is 4 times that of the original cube. By what factor is the volume of the cube increased?

Answer

8

Solution

Note that when the surface area of the cube is multiplied by 4, the sides must be multiplied by 2 since:

$$4 \times 2 \times (lw + lh + wh) = 2 \times (2l \times 2w + 2l \times 2h + 2w \times 2h).$$

Since the sides are doubled, we observe that:

$$2l \times 2w \times 2h = 8lwh.$$

Therefore, the volume is multiplied by a factor of 8.

Problem 11.24 The mineral pyrite is commonly known as "fool's gold" due to its deceptive appearance!

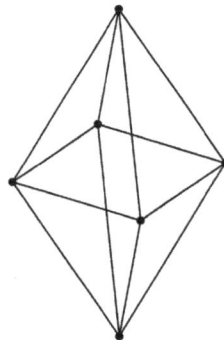

In your quest, you have uncovered a gem-shaped pyrite mineral containing a square-base located at the middle of the gem as shown above. As you fiddle with your newly uncovered gem, you note that the gem is measured to be 8 cm tall and 6 cm wide. If you plan on painting the entire gem gold, how much of the surface will be painted by gold paint?

Answer

120cm^2

Solution

We are interested in determining the surface area of the gem, which can be obtained by determining the area of 8 triangles. Note that the height of the triangle is determined by using Pythagorean theorem. Specifically,

$$h = \sqrt{3^2 + 4^2} = 5$$

Therefore, the surface area of the gem is:

$$8 \times \frac{1}{2} \times 6 \times 5 = 120 \text{ cm}^2.$$

Problem 11.25 **A box shaped treasure chest has outer dimensions 15 cm by 15 cm by 10 cm and the chest is 2 cm thick on all sides. Suppose you have seven dragon balls, each with diameter 5 cm, and you are interested in putting all of the dragon balls in this chest to keep them away from your enemies. Is the volume of the chest greater than the volume of the 7 dragon balls combined?**

Answer

Yes

Solution

Since the chest is 2 cm thick, the amount of space you have for the dragon balls is determined by computing the following:

$$(15 - 2 - 2) \times (15 - 2 - 2) \times (10 - 2 - 2) = 11 \times 11 \times 6 = 726 \text{cm}^3$$

Additionally since each dragonball has diameter 5 cm, the radius of the dragonball is 2.5 cm, and the volume of each dragonball is:

$$\frac{4}{3}\pi \times (2.5)^3 = \frac{125\pi}{6} \text{cm}^3$$

Since there are 7 of them, the total volume of the dragon balls is equal to

$$\frac{875\pi}{6} < \frac{875}{6} \times 4 = \frac{1750}{3} \approx 583.33 \text{ cm}^2$$

Therefore the volume of the chest is large enough to hold the dragon balls. Note, however, we have not worried about arranging the dragon balls. In fact, if you try to arrange them in the chest, they will not fit!